Science for Sustainable Societies

Series Editor
Kazuhiko Takeuchi, Institute for Global Environmental Strategies, Institute for Future Initiatives, The University of Tokyo, Tokyo, Japan

Advisory Editor
Joanne M. Kauffman, Paris, France

Subline Advisory Editors
Hiroshi Komiyama, Tokyo, Japan
Sander Van der Leeuw, Arizona, USA
James Buizer, Arizona, USA
Anantha Duraiappah, New Delhi, India
Thomas Elmqvist, Stockholm, Sweden
Ken Fukushi, Tokyo, Japan
Obijiofor Aginam, Ontario, Canada
Osamu Saito, Kanagawa, Japan
Leena Srivastava, New Delhi, India
Jim Falk, Melbourne, Australia
Rajib Shaw, Tokyo, Japan
Cherry Murray, Arizona, USA
Xin Zhou, kanagawa, Japan
Mikiko Kainuma, Kanagawa, Japan
Annamaria Di Fabio, Florence, Italy
Shiqiu Zhang, Beijing, China
Maike Hamann, Stellenbosch, South Africa
Andra Ioana Horcea-Milcu, Kassel, Germany
Alexander G. Flor, Quezon, Philippines
Noé Aguila Rivera, Vera Cruz, Mexico
Jesús M. Siqueiros-García, Coyoacán, Mexico

This series aims to provide timely coverage of results of research conducted in accordance with the principles of sustainability science to address impediments to achieving sustainable societies – that is, societies that are low carbon emitters, that live in harmony with nature, and that promote the recycling and re-use of natural resources. Books in the series also address innovative means of advancing sustainability science itself in the development of both research and education models. The overall goal of the series is to contribute to the development of sustainability science and to its promotion at research institutions worldwide, with a view to furthering knowledge and overcoming the limitations of traditional discipline-based research to address complex problems that afflict humanity and now seem intractable. Books published in this series will be solicited from scholars working across academic disciplines to address challenges to sustainable development in all areas of human endeavors. This is an official book series of the Integrated Research System for Sustainability Science (IR3S) of the University of Tokyo and the Institute for Global Environmental Strategies (IGES).

Rajendra K. Bera

The Evolution of Knowledge

Scientific Theories for a Sustainable Society

Rajendra K. Bera
Acadinnet Scientific
Bengaluru, Karnataka, India

ISSN 2197-7348　　　　　　ISSN 2197-7356 (electronic)
Science for Sustainable Societies
ISBN 978-981-99-9345-1　　　ISBN 978-981-99-9346-8 (eBook)
https://doi.org/10.1007/978-981-99-9346-8

© The Editor(s) (if applicable) and The Author(s), under exclusive license to Springer Nature Singapore Pte Ltd. 2024

This work is subject to copyright. All rights are solely and exclusively licensed by the Publisher, whether the whole or part of the material is concerned, specifically the rights of translation, reprinting, reuse of illustrations, recitation, broadcasting, reproduction on microfilms or in any other physical way, and transmission or information storage and retrieval, electronic adaptation, computer software, or by similar or dissimilar methodology now known or hereafter developed.
The use of general descriptive names, registered names, trademarks, service marks, etc. in this publication does not imply, even in the absence of a specific statement, that such names are exempt from the relevant protective laws and regulations and therefore free for general use.
The publisher, the authors and the editors are safe to assume that the advice and information in this book are believed to be true and accurate at the date of publication. Neither the publisher nor the authors or the editors give a warranty, expressed or implied, with respect to the material contained herein or for any errors or omissions that may have been made. The publisher remains neutral with regard to jurisdictional claims in published maps and institutional affiliations.

This Springer imprint is published by the registered company Springer Nature Singapore Pte Ltd.
The registered company address is: 152 Beach Road, #21-01/04 Gateway East, Singapore 189721, Singapore

Paper in this product is recyclable.

"To Mansi Balsari, a brave teenager facing a brave new world with dignity and poise."

Preface

What is it about their research that people like Galileo Galilei, Isaac Newton, Albert Einstein, James Maxwell, Max Planck, and hundreds of others seek, not wealth but fame, by discovering some nugget of fundamental knowledge about the Mother Nature? I believe it is sheer curiosity. Their discoveries have changed the face of human civilization through development of new technologies. But there was a time not so long ago when even scientists thought their quest for scientific knowledge was about to end, and scientists would have to move on to other things in life.

In 1894, the American physicist Albert Michelson said, "The more important fundamental laws and facts of physical science have all been discovered, and these are now so firmly established that the possibility of their ever being supplanted in consequence of new discoveries is exceedingly remote… Our future discoveries must be looked for in the sixth place of decimals."

In 1895, Lord Kelvin (William Thomson, 1824–1907) had confidently said, "Heavier-than-air flying machines are impossible" (He had perhaps never seen a flying bird or a bat!). I was studying to be an aeronautical engineer and was obviously appalled when I first heard of it as a student and was trying to understand Kelvin's theorems in fluid dynamics. Such is life! Fortunately for me, in 1903, the Wright brothers (mere bicycle mechanics, but highly gifted with imagination) flew a heavier-than-air powered aircraft they had designed and built. They aptly called it the "flyer". The rest, as they say, is history, and I fulfilled my dream of becoming an aeronautical engineer and a pilot.

In 1905, Albert Einstein showed that space and time are interlinked, and the theory of relativity was born, and an equivalence between mass and energy was shown. The detonation of the first atom bomb (16 July 1945) amply verified that. In 1916, he showed that the presence of matter warps space and time. Physicist John Wheeler pithily remarked, "Matter tells space how to curve, and space tells matter how to move." Our notions of space-time till then were shattered.

The world's first, general-purpose digital computer called Electronic Numerical Integrator and Computer (ENIAC) was introduced on 15 February 1946, under a contract for the U.S. Army and built by John Mauchly and J. Presper Eckert. It occupied about 1800 square feet, used about 18,000 vacuum tubes, and weighed almost

50 tons. It calculated about a thousand times faster than a human. In comparison, seven decades later, the smartphone that sits in our pocket is a piece of science fiction come true.

In 1956, John McCarthy, Marvin Minsky, Nathaniel Rochester, and Claude Shannon proposed the Dartmouth Summer Research Project on Artificial Intelligence and thus was born artificial intelligence (AI). Through several ups and downs and sarcastic barbs since, AI now holds centerstage. The future of human civilization now depends on it, and so does the livelihood of the undereducated. Makes you wonder, what is your worth in the job market if your competitor turns out to be an AI-embedded cognitive robot.

From hunter gatherer to farmer to scientific researcher, to modern-day inventor, humans have bounded from one breathtaking milestone to another. Its success and extent are such that we began using natural resources faster than nature could recycle and replenish. We also started educating people on an increasingly larger scale in science, technology, engineering, and mathematics (STEM) to keep pace with the expanding potential of industrialization, first locally then globally.

Industrialization created so much wealth that the world was suddenly urbanizing at an unimagined scale. Towns and cities were being populated by enthusiastic, rote educated young people who began to form a rapidly expanding middle class. Even more interestingly, the hitherto inherited rich were being brusquely pushed aside by the nouveau riche whose wealth was sourced on their education and their productive efforts in wealth creation. The brainiest among them eventually began to dream of AI. Their creations first excelled in arithmetic and now they endeavor to compete against university educated scientists, engineers, doctors, lawyers, etc. This is turning out to be a grand example of a self-referential system where humans by using their natural intelligence create machines with superhuman intelligence and the intelligent machines through a feedback loop enhance the natural intelligence of humans and so on. Eventually a stage may come, where we preempt natural evolution and self-evolve ourselves into a super species. The present human mind boggles.

Inter alia, the average *Homo sapiens* of today is gripped with fear of its inability to compete against AI machines which can handily uproot them from their present livelihood and make them destitute. Additionally, gifted inventors too can be turned useless once AI machines begin to prolifically invent and discover patentable artifacts and thereby demolish the present patent system. This book ponders over such issues to understand the scale of the problem the *Homo sapiens* face and what they may do to survive. Briefly, the book covers the following topics:

(1) It discusses the uncanny similarity between evolutionary biology and evolutionary knowledge. Darwin's theory of evolution teaches us about adaptation for survival (goal-directed) and about establishing a relationship relative to some feature of environmental order (relational quality). The evolutionary aspect of knowledge is about learning to survive with a better understanding of nature by using intelligence and rational arguments to enable us to anticipate the future and adapt accordingly. In this regard, we depend on a tenuous link between instinct and intelligence. Instinct appears to serve as an invisible hand of nature that

nudges us toward what we must know but what we know is what nature permits. When we observe the world inquisitively, instinct provides subtle cues for further exploration. Intelligence embellishes those cues with one or more plausible conjectures and sifts them through a rigorous fitness trial on available observations by making strenuous refutations to find flaws in every selected conjecture. Surviving conjectures are then used to make predictions of unseen but in principle experimentally observable phenomena that may exist. Conjectures that excel in predictions gain currency and are lauded.

(2) Scientific theories are essentially intellectual constructs. They have advanced at an exponential rate since the industrial revolution (1760–1840). *Homo sapiens* are now experiencing the birth pangs of a new era that encompasses AI, robotics, and automation. They are changing industry dynamics, socio-economic fundamentals, and what it means to compete. The central lesson we learn is that a rapid rate of progress or too much connectivity comes at a price one may not always want to pay or even know how to manage.

(3) Our future employability and survivability will depend on our ability to competitively coexist with AI-embedded machines in the job market. We have to be smarter than AI machines. The alternative is working in the gig economy or finding a rare benevolent benefactor. Darwin's theory of evolution says that our existence depends on how nature selectively weeds out the unfit in a given environment. Progressively, we have thus arrived at a stage where survival dominantly favors those with superior intelligence and the ability to create new knowledge. At every stage of human evolution—hunter-gatherer, agriculturist, industrialist—survival demands progressively greater intellectual contributions and competitively productive skills from individuals for success and a dignified place in society. The time has now come when survival will demand even greater intellectual contributions from individuals which rote education cannot provide, because it is mechanizable in terms of AI. Our future adversaries in the job market will be intelligent machines, other egotistical intelligent *Homo sapiens*, and combinations of them. The heart of AI is algorithmic computation. Computation is all about addition, subtraction, multiplication, division, and comparison of numbers, and problem-solving is all about attaching meaning to numbers. When it comes to knowledge and employability, rote education is now irrelevant.

(4) The exponential rise of science, technology, engineering, and mathematics (STEM), since the 1900s, has completely changed the socio-economic context in which the Patent Act of 1790 and its successive amended versions were enacted. Since then, a person of ordinary skill in the arts (PHOSITA) and in relation to this hypothetical person, the meaning of utility, novelty, non-obviousness, of an invention requiring human ingenuity and the manner in which the invention is to be disclosed to the public in exchange for a limited period monopoly over the invention by the inventor has undergone a sea change. In the last few decades, the world has seen a dramatic change in socio-economic-political structures, remarkable advances in STEM, for example, in information and computing technologies, quantum computing, genetic engineering and synthetic biology,

AI, etc. These have had an enormous impact on the environment in which the *Homo sapiens* find themselves in. Such drastic changes are harbingers of natural speciation, an event that may not be too far off with unknown consequences. The species that succeed the *Homo sapiens* will likely be so far superior in intellect, intuition, and serendipity as to drive the *Homo sapiens* to extinction. This book assumes such an unfolding scenario and, therefore, suggests interim changes to the patent system so that the present debilitating stresses it faces, especially in the form of litigation, are substantially reduced. Our successor species will then perhaps remember us based not solely on our fossil record but also on our ability to anticipate the future and prepare for it intelligently.

(5) This brings us to the vulnerabilities of the present patent systems around the world, in particular, inventions related to advances in quantum computing, synthetic biology, and AI. They have begun to raise serious concerns. Advances in AI are particularly problematic because their influence will be felt on all hitherto patent eligible inventions. Because AI machines have the potential to prolifically invent patentable technology, it will undoubtedly shake the very foundation on which the patent system rests. It will require us to redefine what we mean by novelty, non-obviousness, and written description of the invention (e.g., shouldn't a binary string suffice as written description because it is the lingua franca of computers).

This book in a way chronicles my journey in search of knowledge from a middle-class birth to the fringes of the rich class, initially propelled by rote education but eventually as a researcher and an inventor, and in a modest way as a contributor to AI. The impulse to write this book arose when COVID-19 suddenly burst on the scene and catalyzed an upheaval where rote education must increasingly surrender to AI and humans must reorganize their lives to survive in a world where vast numbers from the middle class must henceforth become gig workers and lose their secure livelihoods to intelligent machines even as they fight against COVID-19 for their lives and livelihoods. We live in dangerous times but also in extraordinary times of human intellectual achievements. I wonder if Charles Dickens would view our predicament as he did of the time, he described in his famous novel *A Tale of Two Cities*:

> It was the best of times, it was the worst of times, it was the age of wisdom, it was the age of foolishness, it was the epoch of belief, it was the epoch of incredulity, it was the season of Light, it was the season of Darkness, it was the spring of hope, it was the winter of despair, we had everything before us, we had nothing before us, we were all going direct to the Heaven, we were all going direct the other way …

Bengaluru, India   Rajendra K. Bera

Acknowledgement

I am immensely grateful to Shobha and Hiten Balsari, who created a serene atmosphere around me that allowed me to write this book. To Sunish Raj, who unobtrusively and gracefully ensured that my computing and other scientific paraphernalia were available, and most importantly the time he spent on discussions on the subject matter of the book.

A Note for the Readers

This book is the result of a study I have been pursuing since the past several years about intelligence and whether it can be mechanized. It has resulted in a series of documents. A selection from those were placed in the public domain and are listed below. The material included in this book is mostly derived from those documents.

SSRN (Social Sciences Research Network)

1. Bera, Rajendra Kumar. On Scientific Theories and Their Impact on Society (August 27, 2021). Available at SSRN: https://ssrn.com/abstract=3912391
2. Bera, Rajendra Kumar. The Evolutionary Nature of Knowledge (August 23, 2021). Available at SSRN: https://ssrn.com/abstract=3909890 or http://dx.doi.org/10.2139/ssrn.3909890
3. Bera, Rajendra Kumar. Knowledge and Employability: The Futility of Rote Education (August 20, 2021). Available at SSRN: https://ssrn.com/abstract=3908420 or http://dx.doi.org/10.2139/ssrn.3908420
4. Bera, Rajendra Kumar. COVID-19 Viewed from a Different Lens (August 10, 2021). Available at SSRN: https://ssrn.com/abstract=3902583 or http://dx.doi.org/10.2139/ssrn.3902583
5. Bera, Rajendra Kumar. AI powered society (September 28, 2018). Available at SSRN: https://ssrn.com/abstract=3256873
6. Bera, Rajendra Kumar. Patent Examination Reforms (January 13, 2017). Available at SSRN: https://ssrn.com/abstract=2898819 or http://dx.doi.org/10.2139/ssrn.2898819
7. Bera, Rajendra Kumar. Patent Subject Matter Eligibility (December 11, 2016). Available at SSRN: https://ssrn.com/abstract=2883838 or http://dx.doi.org/10.2139/ssrn.2883838
8. Bera, Rajendra Kumar. Rethinking Patentable Subject Matter and Related Issues (December 4, 2015). Available at SSRN: https://ssrn.com/abstract=2699219 or http://dx.doi.org/10.2139/ssrn.2699219

9. Bera, Rajendra Kumar. Reforming the Patent System for the Post-Industrial Economy (September 22, 2015). Available at SSRN: https://ssrn.com/abstract=2664035 or http://dx.doi.org/10.2139/ssrn.2664035
10. Bera, Rajendra Kumar. A Minefield of Patents (July 14, 2015). Available at SSRN: https://ssrn.com/abstract=2630681 or http://dx.doi.org/10.2139/ssrn.2630681
11. Bera, Rajendra Kumar. A Rethink on the Expansive Scope of the Doctrine of Equivalents in U.S. Patent Law (May 30, 2015). Available at SSRN: https://ssrn.com/abstract=2612300 or http://dx.doi.org/10.2139/ssrn.2612300
12. Bera, Rajendra Kumar. How Valid are Judicial Exceptions in Subject Matter Eligibility in U.S. Patent Law? (May 10, 2015). Available at SSRN: https://ssrn.com/abstract=2604737 or http://dx.doi.org/10.2139/ssrn.2604737
13. Bera, Rajendra Kumar. Intellectual Property Rights: The New Wealth of Nations (March 3, 2015). Available at SSRN: https://ssrn.com/abstract=2572850 or http://dx.doi.org/10.2139/ssrn.2572850
14. Bera, Rajendra Kumar. Standard-Essential Patents (SEPs) and 'Fair, Reasonable and Non-Discriminatory' (FRAND) Licensing (January 29, 2015). Available at SSRN: https://ssrn.com/abstract=2557390 or http://dx.doi.org/10.2139/ssrn.2557390

USPTO (United States Patent and Trademark Office)

1. Bera, R. K. Patent subject matter eligibility. A response to Notice of Roundtables and Request for Comments Related to Patent Subject Matter Eligibility. USPTO, Federal Register, 17 October 2016, pp. 71485–71489 (5 pages). Submitted 11 December 2016. https://www.uspto.gov/sites/default/files/documents/RT2%20comments%20Rajendra%20K%20Bera.pdf
2. Bera, R. K. Patent examination reforms. A response to Request for Comments on Examination Time Goals USPTO, Federal Register, 25 October 2016, pp. 73383-73384 (2 pages). Submitted 13 January 2017. https://www.uspto.gov/sites/default/files/documents/etacomment_f_bera_13jan2017.pdf
3. Bera, R. K. Patenting Artificial Intelligence Inventions. A response to Request for Comments on Patenting Artificial Intelligence Inventions. USPTO, Federal Register, 27 August 2019, pp. 44889 (1 page). Submitted 21 October 2019. https://www.uspto.gov/sites/default/files/documents/Rajendra-Bera_RFC-84-FR-448809.pdf
4. Bera, R. K. Patent Eligible Jurisprudence Study (2021). A response to the USPTO's request for comments with reference to Patent Eligibility Jurisprudence Study. Notice 86 FR 36257–36260 (4 pages), 09 July 2021. https://www.regulations.gov/search/comment?filter=ksy-99dk-hnd3

Book Chapter

1. Bera, R. K. Biotechnology Patents: Safeguarding Human Health. Book chapter in Innovations in Biotechnology. Eddy C. Agbo, (Ed.), InTech, ISBN 978-953-51-0096-6, Chapter 15, pp 349–376. (2012). https://www.intechopen.com/books/2040
2. Bera, R. K. Synthetic Biology and Intellectual Property Rights, in Biotechnology (Deniz Ekinci, ed), InTech, ISBN 978-953-51-2040-7, Chapter 9, pp 195–232. (2015) https://www.intechopen.com/books/4529
3. Bera, R. K. Synthetic Biology, Artificial Intelligence, and Quantum Computing. Book Chapter in Synthetic Biology—New Interdisciplinary Science. Madan L. Nagpal, Oana-Maria Boldura, Cornel Baltă and Shymaa Enany (Eds.), IntechOpen, 2019. DOI: 10.5772/intechopen.83434. https://www.intechopen.com/books/synthetic-biology-new-interdisciplinary-science/synthetic-biology-artificial-intelligence-and-quantum-computing

Contents

1	**The Evolutionary Nature of Knowledge**	1
	1.1 Introduction	1
	1.2 The Loosening Grip of Religion	2
	1.3 The Rise of Rationalism	8
	1.4 AI-Induced Societal Disruption	9
	1.5 What Is Intelligence?	10
	1.6 *Homo sapiens* Are Driven by Emotion, Not Intelligence	14
	1.7 True Intelligence Seeks Rationality	18
	1.8 Money is Abstract	21
	1.9 Intelligence Versus Artificial Intelligence	24
	1.9.1 The Path from Data to Wisdom	24
	1.9.2 Imitation Versus Innovation	26
	References	29
2	**Scientific Theories Are Intellectual Constructs**	33
	2.1 The Search for Scientific Theories	33
	2.1.1 Theories Should Be Lean and Mean	34
	2.1.2 How Science Progresses	34
	2.1.3 The Web and the Mind	35
	2.1.4 On Building Theories	36
	2.2 An Obsession with Symmetry	38
	2.2.1 What is Symmetry in Mathematics?	38
	2.2.2 Inertial Reference Frame	39
	2.2.3 Maxwell's Equations of Electromagnetism	40
	2.2.4 Nöther's Theorem	42
	2.2.5 Symmetry in Quantum Mechanics	45
	2.2.6 Matter and Anti-matter	46
	2.2.7 Mirror Symmetry in Molecular Biology	46
	2.3 Arithmetization of Cell Theory	47

		2.4	AI and Its Environment	49
		2.4.1	The Brain	49
		2.4.2	Growing Human Population	54
		2.4.3	Exponentially Accelerating Technologies	61
		2.4.4	Know the Power of Your AI Competitor	61
		2.4.5	Science Concepts that Presently Drive AI	65
	2.5		Message for Those About to Step Out of College	65
	References			68
3	**Knowledge and Employability: The Futility of Rote Education**			73
	3.1		Our Evolutionary Trajectory	73
	3.2		How Minds Acquire Knowledge	81
	3.3		Universal Turing Machine	84
		3.3.1	Binary Strings	85
		3.3.2	What is the Central Activity of a UTM?	86
		3.3.3	Capturing Concepts Typographically	89
		3.3.4	Form and Meaning	95
	3.4		Futility of Rote Education	103
		3.4.1	Exponential Growth is Our Undoing	104
		3.4.2	Knowledge Acquisition Requires an Environment	105
	References			106
4	**Evolving AI Raises Human Creativity Concern**			109
	4.1		Introduction	110
	4.2		How Early Inventions Advanced Human Civilization	113
	4.3		When STEM Changed the Face of Man-Made Inventions	117
	4.4		Beyond the Data-Driven World	120
	4.5		Speciation of the *Homo sapiens*	122
	References			122
5	**Vulnerabilities of the Patent System**			125
	5.1		Stressed out Patent System	125
		5.1.1	Since Galileo, the Inventor and Physicist	126
		5.1.2	Knowledge Explosion Since the 20th Century	128
		5.1.3	Algorithmically Designed Biological Inventions	132
		5.1.4	Molecular Biology is Mathematical	133
	5.2		Quantum Physics	135
	5.3		Confluence of AI, Synthetic Biology, and Quantum Computing	136
		5.3.1	Knowledge Integration by Concepts	136
		5.3.2	Integrating the Triad: Mechanization of Speciation	138
	5.4		When Machines Invent, Are Patents Relevant?	139
		5.4.1	Fundamental Tests of Patentability	143
		5.4.2	The Superfluous PHOSITA	144
		5.4.3	Drafting of Patent Applications	144

5.5	Recommendations for the Immediate Future	146
	5.5.1 Patent Validation Board	148
5.6	Conclusions	148
References		149

List of Figures

Fig. 1.1	Major religious groups in the world (2010)	3
Fig. 1.2	Sacrifice of Isaac	16
Fig. 2.1	Agree to disagree	37
Fig. 2.2	Examples of mirror symmetry in biology	47
Fig. 2.3	Four lobes of the neocortex	52
Fig. 2.4	Evolutionary history of the brain	52
Fig. 2.5	Population of the Earth	54
Fig. 2.6	Exponential growth and phase transformation	62
Fig. 2.7	120 years of Moore's law	62
Fig. 3.1	Aerial leaps	75
Fig. 3.2	Recent evolutionary history of *Homo sapiens*	77
Fig. 3.3	**a** World population growth from 1700 to 2100. **b** History of pandemics	78
Fig. 3.4	Euclid of Alexandria	82
Fig. 3.5	Three randomly chosen theorems from Euclidean geometry	83
Fig. 3.6	Turing machine	85
Fig. 3.7	The digital Universe	89
Fig. 3.8	Form and meaning	95
Fig. 3.9	Isomorphism between electrical and mechanical systems	96
Fig. 3.10	Ambiguity. In real life we are often faced with the problem of choosing a point of view	100
Fig. 3.11	The exponential	104

List of Tables

Table 1.1	Isomorphism between atmospheric dynamics and brain dynamics	12
Table 2.1	Arithmetization of cell theory	49
Table 2.2	My personal list of intelligent people who introduced concepts that drive intelligence	66
Table 3.1	Geological timeline	76
Table 3.2	Isomorphism between formal and computational systems	83
Table 3.3	Short biographies of famous people	87
Table 3.4	Primary school arithmetic	91
Table 3.5	College-level arithmetic	92
Table 3.6	Isomorphism between axiomatic (formal) and other systems	97

Chapter 1
The Evolutionary Nature of Knowledge

Abstract There is uncanny similarity between evolutionary biology and evolutionary knowledge. Darwin's theory of evolution is the most comprehensive and successful theory of existence yet conceived. It is about adaptation for survival (goal-directed) and about establishing a relationship relative to some feature of environmental order (relational quality). The evolutionary aspect of knowledge is about learning to survive with a better understanding of Nature using intelligence and rational arguments to enable us to anticipate the future and adapt accordingly. In this we depend on a tenuous link between instinct and intelligence. Instinct appears to serve as an invisible hand of Nature that nudges us towards what we must know but what we know is what Nature permits. When we observe the world inquisitively, instinct provides subtle cues for further exploration. Intelligence embellishes those cues with one or more plausible conjectures and sifts them through a rigorous fitness trial on available observations by making strenuous refutations to find flaws in every selected conjecture. Surviving conjectures are then used to make predictions of unseen but in principle experimentally observable phenomena that may exist. Conjectures that excel in predictions gain currency and are lauded.

Keywords Artificial intelligence · Knowledge · Survival · Rationalism · Evolution

1.1 Introduction

There is uncanny similarity between evolutionary biology and evolutionary knowledge. Darwin's theory of evolution is the most comprehensive and successful theory of existence yet conceived. It is about adaptation for survival (goal-directed) and about establishing a relationship relative to some feature of environmental order (relational quality).[1] Our intelligence is a rational extension of our instinct. It appears fair to assume that our capacity for imbibing and creating knowledge is deeply related

---

[1] Plotkin (1997).

to our biology and therefore to such other things as our ability to network by developing languages, a universal ability to reason, develop culture, and share emotions. Darwin's theory of biological evolution and 'evolutionary epistemology' are analogous, i.e., all knowledge is adaptation, and all adaptation is a distinct form of knowledge. I believe sense-based knowledge anchors mind-based knowledge, and the latter selectively seeks out like-minded individuals of high intelligence rather than accede to the 'knowledge' constructs of cultures.

In my quest to understand AI, I have not found it necessary to deeply study the history of science and technology even though the history is fascinating and includes the evolution of writing, the emergence of science, the scientific revolution, mathematics as the lingua franca of science, the Industrial Revolution, the globalization of knowledge, and the profound impact of science and technology on human affairs.[2] That is because knowledge is advancing so rapidly that for working scientists, delving into the past means taking their eyes off from the future where new adventures lie. However, my understanding of such matters, in brief, is presented in Chap. 5 of this book.

1.2 The Loosening Grip of Religion

There is something odd about God. If he is indeed the Creator, then creation is about the only thing he has done and moved on to other things leaving his creations to manage on their own. Surely humans cannot manage unless they have free will. But if they have free will then they will also be free to do things which God may not like, e.g., conspire against Him. So man is left to figure out how the universe works and what is possible and impossible in our universe. To err is after all a part of being ignorant humans. To learn from one's mistakes and try again is also human. And that is the credo of science; it also includes being curious about and investigating the origin of the universe and how God achieved infallibility. And being infallible, how and why he created all men to be fallible without exception?

In a global population exceeding 8 billion (estimate in December 2022), there is enormous diversity in religion and language. There are more than 4000 religions (divided into churches, denominations, congregations, religious bodies, faith groups, tribes, cultures, and movements) and more than 6800 living languages spoken somewhere in the world.[3] Among the religions, the five most influential are: as in 2010, Christians (31.5%), Muslims (23.2%), unaffiliated, Hindus (16.3%), and Buddhists (7.1%) (see Fig. 1.1). The first two have captured the spiritual imagination of more than half the world's population. They are two of the three Abrahamic religions (the third are the Jews, who throughout their existence have been running from pillar to

---

[2] The interested reader may read Renn (2020).

[3] Rough estimates drawn from Juan (2006) and other sources.

1.2 The Loosening Grip of Religion

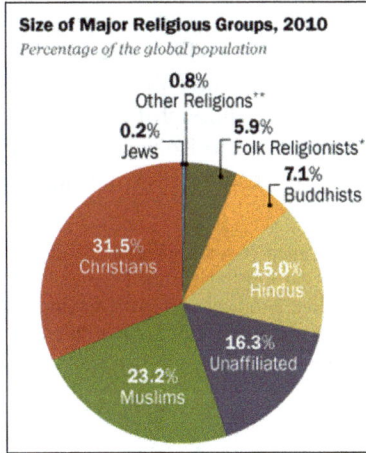

Fig. 1.1 Major religious groups in the world (2010). *Source of figure and quoted text* The Global Religious Landscape. Pew Research Center. 18 December 2012. https://www.pewforum.org/2012/12/18/global-religious-landscape-exec/. Full report at https://assets.pewresearch.org/wp-content/uploads/sites/11/2014/01/global-religion-full.pdf

post to find a piece of land they can call their permanent home[4]) with some common prophets, texts, etc., yet they are perhaps the most divisive among themselves! The reason for their divisiveness (and similarly in all other religions) is easy to find. None has an axiomatic system that guides them. They are guided by an internally inconsistent set of beliefs. This allows the high priests of any religion to select a subset from the total beliefs enshrined in the scriptures to prove whatever they preach as truth to their teeming followers. The system is rigged to ensure that the congregation remains gullible and can be punished for heresy if they demand logical consistency within the complete set of beliefs. Religions fragment when the less gullible seek a modicum of consistency in their beliefs.

The hallmark of intelligent behavior is to seek consistency in the set of beliefs they hold and amend the set when inconsistencies arise. This is anathema in any religion with a substantial following. Those who stray are termed heretics, tortured in various ways, and condemned to "hell" after death after enduring hell on Earth. Further, believers of other religions automatically become heretics and hence condemned to an alien hell. Logically, being religious does not give you any respite from hell. Any significant respite mankind has received has come from those who put their faith in building consistent axiomatic systems of knowledge in science, technology, engineering, and mathematics (STEM) and who never dreamt of calling anyone heretic. Galileo (1564–1642) did not call the Pope a heretic because the Pope did not believe in science. He sensibly kept his silence and endured house arrest until

---

[4] Kirsch (2018). "Ever since Judea was crushed by the Roman Empire, the Jews had possessed none of the things that made for the usual history of a nation: territory, sovereignty, power, armies, kings." (Quoting Isaak Markus Jost from his book *General History of the Israelite People* (published 1832).

his death. Galileo's assertion that the Earth revolves around the sun in support of the Copernican theory was deemed by the Church as heresy.[5] Galileo had dared to publish, in 1632, in his book *Dialogue Concerning the Two Chief World Systems*, a work that compelling endorsed the Copernican system. Although he dedicated the book to Pope Urban VIII, a year later he was summoned to Rome for trial by the Inquisition and forced to renounce all belief in Copernican theories in a signed document:

> I, Galileo Galilei, son of the late Vincenzio Galilei of Florence, aged 70 years, tried personally by this court, and kneeling before You, the most Eminent and Reverend Lord Cardinals, Inquisitors-General throughout the Christian Republic against heretical depravity, having before my eyes the Most Holy Gospels, and laying on them my own hands; I swear that I have always believed, I believe now, and with God's help I will in future believe all which the Holy Catholic and Apostolic Church doth hold, preach, and teach.
>
> ...
>
> I also swear and promise to adopt and observe entirely all the penances which have been or may be by this Holy Office imposed on me. And if I contravene any of these said promises, protests, or oaths, (which God forbid!) I submit myself to all the pains and penalties which by the Sacred Canons and other Decrees general and particular are against such offenders imposed and promulgated. So help me God and the Holy Gospels, which I touch with my own hands. I Galileo Galilei aforesaid have abjured, sworn, and promised, and hold myself bound as above; and in token of the truth, with my own hand have subscribed the present schedule of my abjuration, and have recited it word by word. In Rome, at the Convent della Minerva, this 22nd day of June, 1633.[6]

Legend has it that as Galileo rose from kneeling before his inquisitors, he murmured, "e pur, si muove"—"And yet it moves." Galileo is also reverently remembered for his remarkable observation about science, "The book of Nature is written in the language of mathematics."

The Church's utterly false condemnation forced Galileo to recant his discoveries and led to his house arrest for eight years, which culminated with his death at the age of 77. The alternative offered to Galileo was being burned alive at the stake.

Among Galileo's important discoveries was that acceleration is the same for all bodies subject to gravity, so that heavy and light objects should fall at the same speed. In 1971, Apollo 15 astronaut David Scott, standing on the moon that lacks an atmosphere, dropped a hammer and a feather and found that both touched the moon's surface at the same time. It was a dramatic confirmation.[7] Galileo also pioneered the idea that is deemed obvious today that one can test scientific theories by conducting experiments. Aristotle in his days emphatically did not think so. The Catholic Church was an ardent admirer of Aristotle whose teachings included that the Earth was at the center of the universe. The Church nodded in agreement. Claims of miracles are common in every religion, and they all shun experimental verification. Perhaps Galileo's greatest and most penetrating observation was that the universe could be understood only in the language of mathematics.

---

[5] Cowell (1992).

[6] Halsall (1999).

[7] Lynch (2018).

1.2 The Loosening Grip of Religion

More than two centuries after Galileo's death, the great mathematician David Hilbert (1862–1943) with admiration noted: "Galileo was no idiot. Only an idiot could believe that science requires martyrdom—that may be necessary in religion, but in time a scientific result will establish itself."[8]

In a line of "infallible" Popes, after more than 350 years, one picked up the courage to admit fallibility. That was Pope John Paul II, after the Catholic Church had spent 13 years investigating into the Church's condemnation of Galileo in 1633. In his Papal address of 10 November 1979 to the Pontifical Academy of Sciences, Pope John Paul II, believing that the Roman Catholic Church may have judged Galileo unfairly, called for a commission specifically to reopen the case, saying,

> Galileo sensed in his scientific research the presence of the Creator who, stirring in the depths of his spirit, stimulated him, anticipating and assisting his intuitions.[9]

The commission concluded that Galileo should not have been condemned. In 1992, 359 years after the Galileo trial, Pope John Paul II issued an apology, lifting the edict of Inquisition. "The committee decided the Inquisition had acted in good faith, but was wrong."[10] God never intervened during the Inquisition or thereafter to reveal the truth.

Nicolaus Copernicus (1473–1543) who had put forth his heliocentric theory (that Galileo trusted) in his book *De revolutionibus orbium coelestium*—On the Revolutions of the Heavenly Spheres in 1543 that the sun is at rest near the center of the universe, and that the Earth spins on its axis once daily, fortuitously died just as the book was published and thereby possibly escaped the burning stake. Giordano Bruno (1548–1600) who too believed in the Copernicium heliocentric theory and extended it to cosmic pluralism and insisted that the universe is infinite and could have no "center" was not so lucky. Starting 1593, Bruno faced the Inquisition for heresy, in addition, to charges of denial of several core Catholic doctrines, including eternal damnation, the Trinity, the divinity of Christ, the virginity of Mary, and transubstantiation, was found guilty and burnt at the stake in 1600.[11] In the whole of history, such inhuman practices indulged by men of religion wielding the highest authority in the name of God have no parallel in the history of science.

Belief in the supremacy of science over theology among the scientific elite now reigns. The Church no longer has a stranglehold on scientists that it had in the days

---

[8] As quoted in Eves (1972).

[9] PAS (2003), p. XVII.

[10] New Scientist (1992).

[11] See, e.g., TCE (1912). See also: Stanley (2000). "In 1992, 359 years after condemning Galileo as a heretic, the Vatican apologized and admitted the astronomer [Bruno] had a point. So far, however, the Roman Catholic Church is holding the line on Giordano Bruno, a rationalist philosopher who was burned at the stake for heresy 400 years ago today. ... Today, Cardinal Angelo Sodano, the Vatican secretary of state, said the church "regretted" that it had resorted to violence in Bruno's case, but pointed out that Bruno's writing was "incompatible" with Christian thinking, and that he therefore remains a heretic. ... Giordano Bruno, a Dominican priest whose scientific inquiries led him to argue that the universe was infinite and that Catholic teaching was irrational, is a symbol of stubborn defiance in a society better known for its conformity.".

of Copernicus, Bruno, and Galileo. The scriptures and pontiffs no longer decide which scientific theories are acceptable. Science has acquired a universal following greater than any religion. This was driven home in January 2008, when Pope Benedict XVI (successor of Pope John Paul II) had to cancel a planned visit to La Sapienza University in Rome, a prestigious Italian university, to inaugurate the academic year, after a protest by academics and students for his views on Galileo.[12] Mobilized by the physics department, sixty-seven professors at the university, in a signed letter, portrayed the Pope as a backward theologian who put religion before science and demanded that the Pope's visit be canceled, citing a 1990 speech made by Benedict, then the Vatican Cardinal Joseph Ratzinger, in charge of Church doctrine, had described Galileo's 1633 heresy trial as "reasonable and fair". The students staged a sit-in at the offices of the chancellor, declared an "anti-clerical" week, and protested outside the university, carrying banners insisting that the university is a lay institution and the Pope is not welcome. At one student event, the banner at their lunch read: "Knowledge needs neither fathers nor priests".

Today no Pope would dare call a STEM educated person a heretic. And no scientist whose contribution to STEM has been admirable and breathtaking has ever talked of Jihad or Crusade or of burning, crucifying, or decapitating people. Science transcends religion. Since Isaac Newton and his *Principia*,[13] mankind has gradually turned toward science and reason. Now more so with increasing speed as AI advances!

> It was Galileo Galilei who, following the Middle Ages, first quantified the physical world. He measured the motion, frequency, velocity, and duration of everything from falling stones to swinging pendulums (like the chandelier in his cathedral). It was René Descartes who developed many of the fundamental techniques of modern mathematics and gave us the picture of the universe as a Great Machine. It was Isaac Newton who formulated the laws by which the Great Machine runs.
>
> These men struck boldly against the grip of scholasticism, the medieval thought system of the 12th to the 15th centuries. They attempted to place "man" at the center of the stage, or at least back on the stage: to prove to him that he need not be a bystander in a world governed by unfathomable forces. It is perhaps the greatest irony of history that they accomplished just the opposite.[14]

The inglorious religious fanaticism reached its peak in the Renaissance (14th to seventeenth century), a period in European history when art and literature flourished and marked the transition from the Middle Ages to *Modernity*. With Galileo, the age of modern science began. From his astronomical observations of the planets through a simple cylindrical telescope, Galileo's mind saw the profundity of the idea that we do not occupy the center of the universe, first discovered by Nicolaus Copernicus (1473–1543) some fifty years earlier. "Galileo's work with the telescope unleashed the notion that ours is a sun-centered solar system and not an Earth-centered solar system."[15] It was a dangerous idea that nearly cost him his life. In the end it cost him

---

[12] BBC (2008). See also: Israely (2008). "... the Pope was also sending a warning to the West that reason itself was suffocating faith and destroying its historical identity.".

[13] Newton (1687).

[14] Zukav (1979), p. 50.

[15] Quoting Derrick Pitts in Zax (2009).

1.2 The Loosening Grip of Religion

his freedom in old age till his death. Galileo did not believe that the scriptures were intended to teach astronomy. At the time, humankind's understanding of physics, including of astronomy, remained under the spell of Aristotle. He decreed that all celestial objects were perfect and immutable spheres that went around our stationary Earth, the center of the universe. The astronomical system had already been neatly laid out in books. The Catholic Church swallowed it lock, stock, and barrel. In frustration, Galileo wrote about his countrymen's penchant for relying exclusively on the books from antiquity for learning:

> So far as I can see, their education consisted in being nourished from infancy on the opinion that philosophizing is and can be nothing but to make a comprehensive survey of the texts of Aristotle, that from divers passages they may quickly collect and throw together a great number of solutions to any proposed problem. They wish never to raise their eyes from those pages—as if this great book of the universe had been written to be read by nobody but Aristotle, and his eyes had been destined to see for all posterity.[16]

This was scornful indeed. In his time, not everyone believed in Galileo's claimed observations of the sky nor did they deign to look at the sky through the telescope. Some who did deign still disbelieved their own eyes. So certain were they of Aristotle's wisdom. The challenge to Aristotle was eventually seen by the Catholic Church as a challenge to its divine authority. Hearsay handed down by Aristotle was to be protected by branding challengers as heretics and, if necessary, burnt at the stake.

The conservative guardians of the established order, till artificial intelligence (AI) began to beat humans in unnerving ways, is well described by Arthur Koestler:

> The inertia of the human mind and its resistance to innovation are most clearly demonstrated not, as one might expect, by the ignorant mass—which is easily swayed once its imagination is caught—but by professionals with a vested interest in tradition and in the monopoly of learning. Innovation is a twofold threat to academic mediocrities: it endangers their oracular authority, and it evokes the deeper fear that their whole, laboriously constructed intellectual edifice might collapse. The academic backwoodsmen have been the curse of genius from Aristarchus to Darwin ... they stretch, a solid and hostile phalanx of pedantic mediocrities, across the centuries.[17]

The AI community needs to show a modicum of Galileo's courage and abandon the belief that all *Homo sapiens* have the divine gift of intelligence. And that AI's achievements should be measured in terms of the achievements of the average *Homo sapiens*. First, we have only a vague, instinctive understanding of what intelligence might be. Intelligence is in dire need of a plausible definition (I attempt one in Sect. 1.9). A telescopic view of human society shows that it is riddled with crevices, pockmarks, unevenness, ragged edges over vast expanses, and on rare occasions with high peaks in a small region in terms of what we might call as intellectual achievements. Out-of-the-box thinking is an extremely rare event[18] in a global population that currently exceeds 8 billion people (as of December 2022). If all people were

---

[16] Drake (1957), pp. 126–127.

[17] As quoted in Brake (2009).

[18] If we ignore the liberal grant of silly patents by the United States Patent and Trademark Office. See, e.g., Bera (2008).

intelligent, the world would be inundated by now with obviously useful ideas and the world would have inched (if not leaped) its way to paradise, especially after Jesus Christ preached peace and harmony more than two thousand years ago. His teachings should have had a dramatic effect on humankind in leading them toward tranquility. Instead it spawned Crusades; witch hunting and witch burning at the stake; divisiveness—protestants, Catholics, and more; pedophilia; and so on. AI's expected success depends on the culture of science and science relies on axiomatic systems.

Stephen Hawking recalled an interesting comment by Pope John Paul II at a cosmology conference at the Vatican. The Pope: "It's OK to study the universe and where it began. But we should not inquire into the beginning itself because that was the moment of creation and the work of God." The Pope did not explain why inquiring into the mind of God would be inappropriate. Hawking later joked that he was glad that the Pope did not realize that he had presented a paper at the conference suggesting how the universe began. Later, to an audience at Hong Cong University of Science and Technology, Hawking quipped, "I didn't fancy the thought of being handed over to the Inquisition like Galileo."[19]

No Pope can win a Nobel Prize in physics without first becoming a heretic. Post-Copernicus, all Catholic institutions teaching, researching, and practicing modern physics and biology, by definition, are heretical.

The AI community's strength lies in its ability to think out-of-the-box and think rationally. We built heavier-than-air machines—the magnificent jumbo jets, lethal supersonic jet fighters, heavy-lift helicopters—by asking: "How can we defy gravity rather than how to flap wings more efficiently?" We neither imitated the birds nor compared our aeronautical achievements against theirs. We studied physics, used buoyancy, invented energetic jet flows and whirling rotors, and using Newton's laws of motion concluded that forward motion could lift certain shaped, rigid wings to great heights. AI researchers have not yet reached this level of out-of-the-box imagination. Indeed they should introspect and ask: "How do we imagine? How do we build mindplanes that lead to flights of imagination!" We need to discover axiomatic systems that propel mindplanes and engage in forward thinking that could lift our imagination to the skies. The goal is to discover those axiomatic systems.

1.3 The Rise of Rationalism

The notion that intelligent humans are ubiquitous is a colossal myth. More than 99 percent of the *Homo sapiens* are loath to use whatever intelligence they have even to acquire knowledge when opportunities exist, much less create it. Unintelligent adversarial behavior among nations is the norm even though each would be better off practicing cooperation and collaboration. Their egotistical selfness holds them back; they prefer mutual destruction over construction. This is raw, untamed human intelligence at play. There is widespread irrationality among *Homo sapiens*; being

---

[19] USA Today (2006).

consistently rational is extremely rare; blatantly rationalizing what cannot be rationalized is extremely common. The only example of genuine rationalism in practice is seen in the scientific community where collaboration and cooperation is the expected norm that cuts across national, religious, and social boundaries.

Intelligence and instinct are separate. There are genetically coded basic instincts, and there are cultivated instincts acquired through learning. Cultivated instincts sharpen our ability to form concepts and build axiomatic systems. A well-developed instinct provides short cuts in decision-making; it is a very powerful tool in the hands of a few gifted people who advance knowledge by the Popperian process of conjectures and refutations.[20] Belief is a glorified conjecture intolerant of refutation, and it demands unquestioned acceptance; an axiom is an inspired conjecture that invites refutation to improve itself or die. An axiomatic system thus allows itself to evolve in a Darwinian sense. An axiomatic system modeling the universe will generally have some features that remain constant (e.g., the total energy in a closed, isolated system; the speed of light, etc.), some features over which we have some degree of control (e.g., the amount of information circulating in a feedback system such as an autopilot executing a certain task), and some features over which we may have no control (e.g., freak weather). In a mathematical representation of an axiomatic system these are generally referred to as constants, parameters, and variables respectively.

The limitations of our brain-mind system generally forces us to build axiomatic systems in layers so that our mind is not overwhelmed. At the lowest level, features either remain constant or approximately so or maintain a reliable relationship. The middle level is dominated by parameters which we can control within certain limits and precision; at the highest level variables may take values over which we may have no control or which may be outside its scope. To mitigate unwanted instabilities and out of scope situations, an AI system must have contextual information and the enveloping boundary within which it can function in terms of the constants, parameters, and variables the system relies on. Contemporary AI systems are likely to develop within this broad framework.

1.4 AI-Induced Societal Disruption

Rapid advances in AI are poised to disrupt society further over and above the disruptions already under way, e.g., by accelerated climate change, pandemics, rising socio-economic inequality, unprecedented political crises sprouting in democracies, public sectors starved of funds, etc. Public confidence in existing political and economic institutions is ebbing. The capitalist system now faces yet another existential crisis. Trump's America First agenda created its own set of political and economic disruptions; his relentless trade war against China caused unforeseen and unanticipated global disruptions; a sudden lethal outbreak of the COVID-19 pandemic has convulsed the world; and Russia's invasion of Ukraine has sent tremors of a possible

---

[20] Popper (1963).

World War III. Worldwide there is a crisis of confidence in governments, in political systems, in business leaders in the private sector, and in mammoth communication and media companies. Unethical handling of fake news, disinformation, privacy concerns, etc. abound. Ethical crises are sprouting everywhere on a massive scale and so is the display of moral turpitude by people holding positions of trust, including those in venerated religious institutions. A pervasive degradation of society is well underway. As Joseph Stiglitz notes:

> The financial sector has been marked by predatory lending, market manipulation, and abusive consumer-credit practices. Automakers have been caught gaming environmental regulations. The food and beverage industry is knowingly contributing to childhood obesity around the world. Pharmaceutical companies are pushing addictive drugs even as they claim otherwise (while eschewing research into desperately needed new antibiotics).[21]

In such a world, AI becomes the only rational, reliable, and viably programmable trusted source of information and advice. On the downside, when the common man in search of a job is vetted by an AI machine, how will he fare? Can he blame the AI machine for nepotism if a machine is favored instead of a human? Or will it mean the destruction of hope?

What will be the place and role of dissent when discontent rises? What role will religious institutions have in society? What will be the structure of society? What will be the size of the gig economy? What is the risk of social catastrophe even though AI decisions are rational but lack empathy? Controlling the general population is a primary role of those wielding power and privilege. Can AI machines be programmed for such a role if its only option is to rule by fiat? If AI takes on the role of elite decision-making bodies, how can humans play any meaningful, democratic role in society? In an AI-dominated society, the common man becomes a member of a bewildered herd, deprived of an audible voice; they will be spectators not participants. Efficient AI-driven society will be run under top-down authoritarian control with the common man treated as an ignorant, disposable burden, devoid of any meaningful role in decision-making.

If deprived of gainful employment, humanity will be deprived of hope and purpose. It will simply wither away and be replaced by a more intelligent species. Can humans ever gain dominance over AI-delivered rule of law? The answer appears to be no. Any such attempt will likely mean destruction of the *Homo sapiens* given their present population size and acute dependence on automation for mass production of goods and delivery of services at affordable prices.

1.5 What Is Intelligence?

In this book we posit that human intelligence is an adaptive feedback controller that comprises two broad categories: (1) axiomatic (governed by a human concocted axiomatic system based on conjectures and refutations) and (2) intuitive (governed

---

[21] Stiglitz (2020).

1.5 What Is Intelligence?

by intuition, gut feeling, premonition, and serendipity). In principle, the axiomatic is mechanizable while the intuitive can be weakly simulated. We further posit both are strictly constrained by some omniscient, omnipresent Laws of Nature of unknown origin, and they are unfathomable. Hence our understanding of these laws is through conjectures, which we continuously and endlessly strive to refine, replace, overthrow, or discard by self- or community-driven refutations. This process is admirably described by Karl Popper in *Conjectures and Refutations*.[22]

The world has about 2668 billionaires with a net wealth of $12.7 trillion,[23] a shocking reminder that creative human intelligence in a global population of 8 billion is barely visible, else why would there be an anomaly where the world's richest 1% have more than twice as much wealth as 6.9 billion people,[24] who live in constant fear of losing their jobs, income, and status is society. Clearly, our perception that every *Homo sapiens* is gifted with intelligence is grossly wrong if the global economy is being run by just a few thousand people.

Since Galileo, only a tiny population of genuinely intelligent people have created scientific knowledge. It was seeded, nurtured, and advanced largely in Western Europe; it was willingly, freely, openly, and altruistically shared with the rest of the world. Yet the eulogized masses deemed intelligent *Homo sapiens* as individuals were unable to grasp the significance of that knowledge, much less imbibe and further it or even exploit it for their own upliftment. The remarkable indifference to imbibing scientific knowledge by the average *Homo sapiens* is an amazing paradox in human history.

A basic characteristic of intelligence is that it seeks to discover organized structures in the universe and an efficient means of indexing them for future recall and dynamically redefine and recategorize them as new knowledge arrives. The term "organized" is used in a recursive sense, i.e., it involves the repeated application of a rule, definition, or procedure to successive results; there can be nesting of structure, i.e., a structure can exist within a structure. Mathematicians are very familiar with recursion, an example of which is the factorial function.[25]

Without being aware and conscious humans cannot be intelligent. The two properties connect us with the universe such that back and forth information flow is established using our sense organs. Our beliefs shape our intelligence, and they are sharpened by continuous feedback loops between ourselves and the rest of the universe. Intelligence as controller enables us to stitch together context and information in an organized way (say, by using formatting rules, precision, brevity, qualifying attributes, etc.) so that information flow can be mechanized, stored, edited, retrieved, copied, deleted, annotated, sorted, etc., and isomorphically connected to a meaning

---

[22] Popper (1963). See also Popper (1934).

[23] The World's Billionaires. Wikipedia. https://en.wikipedia.org/wiki/The_World%27s_Billionaires#2022. Accessed 11 December 2022.

[24] Whiting (2020). There is no sign of disparities closing soon. See, e.g., WIL (2021).

[25] The factorial of a number n is defined as $n! = n \times (n-1) \times (n-2) \times \ldots \times 3 \times 2 \times 1$, which has the recursive form $n! = n \times (n-1)!$ with $1! = 1$, and $0! = 1$. The factorial represents the total number of distinct sequences–the permutations–of n distinct objects that are possible.

Table 1.1 Isomorphism between atmospheric dynamics and brain dynamics

Atmospheric dynamics	Brain dynamics
Storm	Mind
Eye of the storm; strange attractor	Axiomatized knowledge base; intelligence
Weather data archive	Knowledge archive
Weather modeling	Discovering an axiomatic system
Unpredictability of weather	Indecisiveness of the mind
The seasons	Knowledge disciplines

in some real or imagined universe. An information carrying message thus prepared can be deciphered and interpreted by a trained human mind.

Human intelligence acts on information in a controlled manner based on its conscious awareness of the environment but above all on its instinctive and intuitive understanding of the world derived from experiences and held beliefs. Higher the intelligence, greater the abstraction level at which it contemplates, assesses, and acts with foresight. If the abstraction level is supported by an axiomatic system, then the belief system can be arithmetized and mechanized and packaged as artificial intelligence. Since information is physical[26] intelligence too must be physical; otherwise it cannot process information. We thus posit the mind is analogous to a storm and intelligence is analogous to the eye of the storm. (See Table 1.1.) A raging storm exists in a physical atmosphere only as a chaotic pattern. Likewise, a hyperactive mind exists as a chaotic pattern in the biochemistry of the brain. The counterpart of the eye of the storm is the rational (axiomatized) knowledge base held in the brain in appropriate biochemical form. What we know as a strange attractor in chaos theory appears as intelligence in brain dynamics. Like the sudden build-up of an unstable atmospheric storm under certain atmospheric conditions, the mind too erupts into chaotic activity under certain biochemical conditions.

All this leads me to believe that as we explore further, additional analogies will be found, e.g., atmospheric temperature could be connected to what we may call a "community temperature" that characterizes intelligence variation in a community that blends rational and intuitive aspects of intelligence along the lines of black body spectrum that defines temperature in physics. From this one can then draw further analogies from thermodynamics. For example, a community's ability to do useful work with other communities running at different "temperatures". Relatively higher community temperature (say, in terms of superior information processing capability of its scientists) would turn the community into an engine of economic growth; a low community temperature would turn the community into a refrigerator for preserving ignorance when it insulates itself from a higher temperature environment (e.g., by becoming religious).

---

[26] Landauer (1991).

1.5 What Is Intelligence?

Although outside the scope of this book, the idea of "community temperature" is worth pursuing since it is now well established that information theory and thermodynamics are related through the insights provided by Calude Shannon[27] and Rolf Landauer.[28] Further, nonlinear systems when they become unstable share certain characteristic features through common mathematical structures. This fact too will provide opportunities to discover isomorphic relationships between atmospheric dynamics and brain dynamics. This raises the possibility that studying atmospheric dynamics may be an easier way to advance AI by drawing analogies rather than through experimentation on brains which is far more tricky.

Human knowledge flows like heat in a thermal environment. For example, community-held knowledge may be viewed as flowing from high community temperature hot spots (e.g., university faculty, researchers, managers, etc.) to low community temperature pools (e.g., students, research assistants, trainees, etc.). Heat flow can be resisted by thermal insulation; likewise knowledge flow can be resisted by belief barriers. Heat dissipation, absorption, and release analogs in knowledge occur according to the absorbing capacity or creativity of the receiver and donor community. The counterpart of heat reservoirs (coal, gas, etc.) is libraries, knowledge archives, experts, etc. The counterpart of heat radiation is spreading of knowledge via electronic means.

The network of *Homo sapiens* resembles an electrical circuit, and electrical circuits can encode algorithms and compute. We can, in principle, find out what human society is computing (or where it is heading or what it is doing) by drawing parallels with social activities, beliefs, and customs. AI does not need to understand awareness, curiosity, contemplation; it only needs the hardware and software means to sense, measure, compute, and deliver. The means in the future will include quantum computers and quantum algorithms. The real challenge yet to be faced: "Can AI make conjectures and build axiomatic systems?" My belief is it can because conjectures are all about making correlations and finding quantum entanglements in Hilbert space. Experience is nothing but the ability to store previously generated information and indexing it for rapid recall.

Intelligence is required to grasp subtle, abstract patterns; to manage change; and to discover paradoxes. This is no mean task. According to the Church-Turing thesis,[29] this means that if we have a rich enough axiomatic system at hand, we can organize structures (social, economic, political, bureaucratic, mechanical, electrical, civil, etc.). It is an irony that while we endeavor to instill human like behavior in AI, we also endeavor to organize ourselves into automatons that eerily resemble axiomatic systems but when interpreted in our Earthly world the fit is vitiated by emotion. What has been achieved was made possible by emphasizing rote education that deemphasizes human intelligence. By such moves toward mechanization we sidestep creativity!

---

[27] Shannon (1948).

[28] Landauer (1991).

[29] It states that a function on the natural numbers can be calculated by an effective method if and only if it is computable by a Turing machine.

Intelligence is about detecting isomorphisms given a context. Its basic characteristics are that it: (1) follows propositional logic (zeroth-order logic), (2) follows Peano arithmetic, (3) makes curiosity-driven conjectures, (4) evaluates and updates conjectures for consistency and greater scope, (5) makes deterministic and probabilistic correlations at multiple levels of abstraction, (6) seeks community affirmation and criticism of adopted conjectures, (7) ferrets around for isomorphisms, (8) imagines and explores unknown worlds, (9) is physical and bound by the laws of Nature, and (10) seeks to build through exploration, varied structures at all scales of space–time and record them for future generations to continue the work seamlessly. Intelligence does not seek truth; it seeks to minimize the information needed to interact with the universe in a self-referential way, that is, by induction.

Einstein summarized his view on how science progresses:

> Science is the attempt to make the chaotic diversity of our sense-experience correspond to a logically uniform system of thought. In this system single experiences must be correlated with the theoretic structure in such a way that the resulting coordination is unique and convincing. The sense-experiences are the given subject-matter. But the theory that shall interpret them is man-made. It is the result of an extremely laborious process of adaptation: hypothetical, never completely final, always subject to question and doubt. The scientific way of forming concepts differs from that which we use in our daily life, not basically, but merely in the more precise definition of concepts and conclusions; more painstaking and systematic choice of experimental material; and greater logical economy. By this last we mean the effort to reduce all concepts and correlations to as few as possible logically independent basic concepts and axioms.[30]

Einstein's tremendous contributions to the theory of relativity and quantum mechanics have radically changed our view of the universe. His insights show how rare is the intelligence that can fathom the secrets of the universe. The world had to wait for a single gifted individual to enlighten us. And yet, only a miniscule fraction of the global population is enlightened!

1.6 *Homo sapiens* Are Driven by Emotion, Not Intelligence

The universe came into existence without any *Homo sapiens* in sight, without living creatures, without any trace of intelligent or emotional beings to marvel at the awesome universe that was emerging, and the Creator apparently did not feel so insecure that he needed *Homo sapiens* around to gawk and eulogize him in utter servility till death wipes him out. When He did create the *Homo sapiens*, the outer edges of the universe were well beyond his sight; He did not inform him of his humble origins that he evolved from a microscopic single-celled creature billions of years ago and that his recent ancestors are the great apes and the chimpanzees are his nearest cousins. Life began with feelings (sensors), evolved to acquire emotions, and finally with the genus Homo dull sparks of intelligence emerged. God's messengers were left totally in the dark of all these happenings; they still remain so.

---

[30] Einstein (1940).

1.6 *Homo sapiens* Are Driven by Emotion, Not Intelligence

Homo sapiens appeared billions of years after its single-celled ancestors showed up on Earth, with absolutely no idea about where, why, and when he was created. But ironically he was laced with emotion, hunger, an instinct for survival and reproduction, fear, and a necessity to commune to find safety in numbers. The brain, when it first appeared in some ancestral species, had no intelligence. The brain essentially auto-controlled the body in its various functions, e.g., progenitor, predator, or prey. Emotions ruled, intelligence (if present) hibernated. When it did awake it suffered from deep inertia, unwilling to exercise and expand its mental horizons beyond its hunting grounds. From lack of use, even today intelligence in humans is seen in trace quantities and is used so sparingly that primitive herd mentality prevails.

The *Homo sapiens* already had body language (built-in as instinct), but with languorous awakening of intelligence, they laboriously developed spoken language, the written language, logic for reasoning, arithmetic for counting and book-keeping, algebra, physics and the sciences, technology, the printing press, and finally mechanized supermassive communication networks. The development of language, computation, and mechanizable communication, the crown jewels of the *Homo sapiens*, came from an amazingly small group of people endowed with superintelligence and it took millennia. Even then hordes of *Homo sapiens* did not rush to acquire knowledge and sharpen their intelligence or hanker for it. They preferred passive ignorance, hearsay, and dogmatism of religion. To find employment the modern *Homo sapiens* grudgingly seek rote education that teaches them to repeat, replicate, and rehearse what pioneering geniuses have created. Even incremental contribution from the rote educated to the storehouse of rational knowledge is rare, finding isomorphic connections between disciplines of knowledge rarer, and creating axiomatic systems extraordinarily rare. Large-scale adoption of rote education is a recent phenomenon driven by the socio-economic opportunities created by the Industrial Revolution (1760–1840). Those aspiring to win Noble Prizes seek joy and beauty in the austerity of scientific knowledge; those seeking salvation (the huge majority) seek succor in the rituals of religion.

Almost all *Homo sapiens* suffer from acute fear fueled by ignorance. Fear induces irrational behavior, loss of self-control, and promotes exploitable dependency on others. *Homo sapiens* have never possessed enough intelligence to confront fear. This has led to the irrational creation of gods and eventually a very powerful monotheist God to be feared, obeyed unconditionally (in the biblical narrative, God asks Abraham to sacrifice his son, Isaac, and Abraham meekly agrees to comply, but God relents! See Fig. 1.2). The God is omnipotent. The layman does not ask, "Can God create a stone so heavy that He cannot lift it?" God is omnipresent and omniscient, yet blind to human miseries. The list is long. An intelligent mind abhors such paradoxes.

Man, heavily preoccupied with defending himself in a world "red in tooth and claw", surely could not have created a rational God so the alternative was to imagine one that could be appeased. If one was destined to be a slave to oppressive Kings and Emperors, one might go a step further and serve an even bigger oppressor who could deal with his present oppressors in their afterlife. Great intelligence was required to douse fire with fire. This singular intelligent feat was achieved by man creating God in his own image, a priestly class in the service of God, and conjure an unimaginable

Fig. 1.2 Sacrifice of Isaac. Abraham en Isaac, by Rembrandt, 1634. https://commons.wikimedia.org/wiki/File:Rembrandt_Abraham_en_Isaac,_1634.jpg

Hell where unrepenting oppressors could spend their afterlife. This God necessarily had to be supernatural, capable of conjuring miracles, unapproachable so that he could be venerated. The power of miracles immediately ensured that God could not be rational and accountable. His being unapproachable meant that one cannot demand anything from him, you can only be a supplicant. These are the natural creations of a mind filled with fear. To gain favors he must supplicate. The final act was to make even Kings and Emperors believe in this imagery. They eventually did. Intelligence was used to create a logical paradox: "Man created God in his own image; God created man in his own image." When I stand in front of a mirror, did I create my image in the mirror or did my image create me. It took some time for Kings and Emperors to learn that the priestly class could match them in the art of oppression that included such things as witch hunting and burning heretics alive at the stake. All religions survive based on such man-made paradoxes.

Thus, when we face a crisis, we do not turn to any Tom, Dick, and Harry because we do not expect to see problem-solving, real intelligent people around but seek out those who are smarter than we are for help or consultation. And even there we choose the best among the accessible. The average *Homo sapiens* gives as much attention to his brain as to his appendix. Only rote education could turn some of them into sufficiently mechanized, "intelligent" humans to make the Industrial Revolution a success and create a prosperous middle class. In the 1950s, some truly intelligent people embarked on creating AI. Now in the 2020s, AI is poised to decimate the middle class by making *Homo sapiens* jobless and economically worthless. There was a time when humans judged and graded AI; now we are to be judged by AI.

Just as we have difficulty defining beauty and evade behind the aphorism "Beauty is in the eye of the beholder", so it is with intelligence. Between the beauty and the beast lies infinite variety and so it is between intelligence and stupidity. Phenomenal beauty is rare and so is genuine intelligence. Between genuine intelligence and stupidity lies a phenomenal range of confrontational mixture of reason and unreason. The wide midrange in the intelligence scale appears as a cacophony of odd perceptions ranging from "A wise man never knows all, only fools think they know everything." The world is overburdened with people who pretend they know almost everything (Jack of all trades, master of none). Wars between "-isms" are a good recurring example. Those dealing with rational science freely admit that they do not know the truth and never hope to know it either. They do not settle disagreements with bloody wars. They let the opposition die a natural death. Religion encourages ignorance, science tries to dispel it. Isaac Newton, James Maxwell, and Albert Einstein were a class apart even among geniuses. Ironically, the public cares far less about science than about their perception of scientists. The public is also generally unaware that scientific knowledge once encapsulated in mathematics is morally neutral. How can it be otherwise? Scriptures and prayers can no longer compete against science and AI-embedded technology.

For our purposes we posit *consciousness* as the state of a normally functioning brain (natural neural network) of a biological entity that is biologically switched on sometime after conception and before birth, analogous to the switching on of an electrical appliance to a power source. On death, it is biologically switched off. Between the on–off states, it can be put into sleep mode when the brain remains partially alert to continue with essential housekeeping tasks needed to keep the body alive and secure. Consciousness is a biochemical regulator that essentially regulates the flow of energy (primarily electromagnetic) in the brain. The brain functions like computer hardware with some primitive operations, e.g., computing, probabilistic decision-making, and information management capabilities hard-wired. The further evolution of the brain is what we call the mind. This evolution is predominately captured in the dynamic neural network connections the brain engages in as it interacts with the environment in an introspective manner (i.e., using self-referential feedback loops). The mind is the brain's software. The topology of the neural network and the brain jointly describe and characterize the evolving biological brain-mind entity in its relationship with the environment.

The body sensors that feed information to the brain are sensory organs (e.g., eyes, ears, nose, etc.). What they detect contributes to awareness. Thus, awareness is multifaceted, biological, and biochemical. The sensors are tunable, i.e., their sensitivity can be calibrated, filtered, amplified, etc. with or without the aid of auxiliary non-biological devices. Our brains respond to inputs that come from awareness by placing and *understanding* the inputs in a context, e.g., by recalling and searching for relevant information from memory, related past experiences, mental repositories of information acquired from training, rote education, etc. It essentially narrows down and prioritizes the contexts in which further analysis can be applied. *Intelligence* analyzes and produces a response as per its capability. That capability in most *Homo sapiens* is highly impaired because it is driven by emotion. *Homo sapiens* rarely go past the understanding stage if it requires them to use their intelligence.

1.7 True Intelligence Seeks Rationality

> Mathematics is a game played according to certain simple rules with meaningless marks on paper. – David Hilbert[31]

AI researchers must bear the following in mind. True intelligence feverishly seeks rationality, consistency, and the limits of attainable knowledge by the Popperian process of making conjectures and refutations. It communicates in the language of mechanizable, axiomatic mathematics. The intellectual *Homo sapiens* are amazingly rare. Creating new, non-obvious algorithms is an intelligent activity, and such activity, in principle, is mechanizable. AI is about creative computational activity, e.g., creation of an axiomatic system, deriving theorems, rational decision-making, detecting paradoxes, etc. Just as information is physical so is intelligence.

An algorithm specifies two things: (1) specific operations to be performed and (2) control statements. Algorithms lie at the heart of computing and software products. Software plus a computer is a means of representing and using knowledge. Splitting the software into program + data is an art. What and how primitive knowledge objects are to be apportioned between hardware and software is also an art. Any software product develops organically like the creation and growth of a human being. It is conceived, takes an initial form through multiplication of some basic units, evolves to a standard form and then conforming to that form it scales in size, and adapts to the environment through content modification and refinement. During its lifetime, at different stages, the stage and its various elements are given names to indicate the state of its development, e.g., software product, program, procedure, etc. Most important software products develop from generation to generation. Some of the important names in vogue are briefly described below. It is assumed that the reader is generally familiar with these terms and with software development in general.

Data types

A data type is an organized, encapsulated list of operands, operators (with implementation details), other existing and accessible data types (implicitly by reference), and the type's relationship with other existing types. Certain primitive data types and their relationships with each other are assumed to exist when creating new data types, e.g., *integer* along with operators plus, minus, multiplication, and division; *character* along with operators that concatenate, delete, copy, etc. The job of an operator is to act upon specified operands and transform them in specified ways to produce an output. If the operator is not provided any operand, it produces a prespecified default

---

[31] As quoted in Rose (1988).

1.7 True Intelligence Seeks Rationality

output. Commonly used data types are *array, stack, queue, linked list, trees, vector, matrix, complex number*, etc.

Very powerful data types can be used to passively create and store knowledge and to introspect. Such introspective data types autonomously introspect when the computer is idle.

Expressions and statements

An expression is a sequence of operators and operands presented in a specified format. The expression encodes a computation to be executed. An expression may result in a value and may create side effects (e.g., trigger some activity).

A statement is a syntactic unit that expresses some action to be carried out, e.g., it may take the form of an assignment statement (e.g., "x = 5"), or a function call (e.g., "do something with (x, y)"), or a block of statements to be executed sequentially.

Branches and loops

A branch coding statement executes an expression which evaluates to 'true' or 'false' and accordingly chooses one from a pair of options. For example, "if (N is less than 5) execute statement 1; else execute statement 2"). A branching structure represents the *if-then-else* style of argument.

A loop coding structure has an entry point, one or more loop exiting condition, a looping body, and a loop exit point. Its generic structure is:

Enter loop {*if* loop exiting condition is not true *then* execute looping body *else* exit loop}.

A loop structure represents an iteration statement. Such a structure comes under the ambit of Turing's halting theorem[32] and provides a formidable limitation to axiomatic reasoning which cannot be overcome by any form of irrational reasoning (e.g., religion). The limitation is our intellectual inability to unerringly find a loop exiting condition under all situations in an axiomatic system.

Procedures

A conveyed procedure is a structured articulation of an organized set of activities meant to reach an end state from an initial state. The organized set of activities usually comprises primitive actions and detection means that we assume can be done without further elaboration by some available means (hardware or software) and organized into a network. This network is the procedure. The network topology defines the complexity of the network and the class of problems the network can deal with subject to Turing's halting theorem (i.e., the specific arrangement by which the network nodes are linked, or messages go back and forth between sender and receiver nodes as a function of time). A network node is a general term; it can represent a primitive entity or activity or various organized arrangements of such primitives, i.e., another network. A network can call itself, i.e., self-reference and therefore may have chaotic behavior built into it without our knowledge! A network may be nested in

---

[32] Turing (1936). Turing proved that there can be no general procedure to decide if a self-contained computer program will eventually halt.

various ways, e.g., by design, by oversight or emerge in action in which case nodes may interact with subsets of nodes it is linked with or with better-connected nodes.[33] Procedures can be chained together or nested within in various logically permitted ways as in mathematics.

A procedure is entered by calling it by its unique *id* and giving it a list of input values to get started with, and a list of placeholders to be populated by the procedure before it exits. A procedure can be autonomous, i.e., it can be called without receiving any inputs or returning any outputs but signaling that it has exited. A procedure can call other procedures and therefore can import or export information with the rest of the world, or even call itself recursively. The input values provided at entry to a procedure serve as its axioms. The procedure using permissible rules for manipulating the input (axioms) produces the output (theorems). If no uncorrectable errors occur between entry and exit the output placeholders carry the theorems the procedure has produced on exit.

Programs and higher-level structures

A program is an autonomous procedure containing within it a set of other procedures it can call. A program is the coordinator of an ensemble of procedures. The distinction is one of size and controlling authority and not of concept. Thus, a standalone procedure is also a program. A program represents a task that is chunked into smaller parts and linked into a network. A chunk at a smaller scale may be a procedure or a group of procedures. There is no unique way to chunk a program. Creativity and artistry is involved, especially if the task is complex (e.g., requiring the program to dynamically reorganize itself during execution), or program size is large, or must be guaranteed not to fail, or has stringent time and resource constraints imposed upon it. Very large programs, e.g., software products are usually modularized where each module is usually organized around a function or specialty or security considerations, etc.

Modular design of any complex system is an art. It forces one to be creative, and creativity always requires intelligence. Modularity determines how easy or difficult it will be to build a knowledge system, insert new knowledge, and modify the system. Highly modular systems are easy to build and manage. If we want to build a theorem-generating system which uses axioms and rules of inference, then we must remember that the "knowledge" of such a program resides implicitly in the axioms and rules and explicitly in the body of theorems which it produces. The modularization must bear in mind how knowledge is intended to be used. For example, are inferences to be made on the fly as information streams in, or extensive analogies and comparisons are to be made before drawing conclusions in which case information redundancy should be minimal, or for human understanding by providing multiple analogies, alternatives, and isomorphisms, or for replication and action according to a dynamic schedule (as Nature does with life forms—DNA for replication and RNA for action by creating proteins)? Software designers face such critical questions for which there are no easy answers.

---

[33] See, e.g., Mariani et al. (2019).

Execution flow

In a group of statements, statements are executed sequentially in the sequence they are presented except when a branch or a loop intervenes to interrupt and redirect the execution flow. If statements comprising a group have no dependencies among themselves during execution, then those statements can be executed in parallel. Likewise, and recursively, for groups comprising a larger group.

Neural nets

Also known as deep computing, it has been hyped so much that it appeared to be a mysterious magic wand that required wizards to wave and do magic. Thankfully that ballyhoo is now rapidly dying once it was discovered that they are merrily hackable.[34] Intelligence does not produce miracles; it produces rational thought and computable algorithms. If you cannot explain why an algorithm works, you are selling snake oil. An algorithm that works must belong to an axiomatic system. Finding that axiomatic system is an intelligent activity.

1.8 Money is Abstract

Economists are amazingly audacious; it appears their goal is to make astrology look respectable with such dubious means as a dress code and the art of obfuscation practiced by psychologists. It is therefore remarkable that they continue to establish themselves, especially in the corridors of government policy making, in positions of intellectual authority, oblivious of their ignorance. The scientific community marvels at their continuing ability to savor unparalleled success without knowing any of the science and technology that drives the modern economy. The social success of economists is second only to those of religions, which too have no understanding of science and technology, leave alone any rational discipline of knowledge. Despite all this, the *intelligent* socio-economic creature, the *Homo sapiens*, sees nothing wrong in entrusting their faith, trust, and future in economists and priests, rather than flocking to the scientists for advice. The obvious reason is that people prefer to remain ignorant by suppressing their intelligence. Suppressing emotions is far tougher than suppressing intelligence; astrology, prejudices, religion, and gambling are intimately close to our basic instincts.

As John Kenneth Galbraith (1908–2006) quipped: "Economics is extremely useful as a form of employment for economists." And "If all else fails, immortality can always be assured by spectacular error." Even after more than a decade of the 2007–2008 financial crisis, economists still remain blind to the fundamental errors they harbor in their thinking.[35] They are still blind to the obvious source of money in

---

[34] Stewart (2019).

[35] For a very frank and unflattering opinion on economists, see Piketty (2014), in the Introduction: The Theoretical and Conceptual Framework. Piketty says, the profession churns out purely theoretical results without even knowing what facts are needed to be explained. "To put it bluntly, the

the modern economy; banks create it out of thin air, simply by making loans since today's money is essentially credit. It works like this:

> … bankers simply wave a magic wand and make the money appear, secure in the confidence that even if they hand a client a credit for $1 million, ultimately the recipient will put it back in the bank again, so that, across the system as a whole, credits and debts will cancel out. Rather than loans being based in deposits, in this view, deposits themselves were the result of loans.[36]

For this to work, the trust the bank places in the client must be absolute and inviolable. Real drama occurs when the loan recipient does not return the loan to the bank. Sometimes they don't.

Another "intellectual" topic economists deal with is inflation. Inflation cannot be controlled by controlling money supply unless it can be precisely timed in precise doses because of the omnipresence of uncertainty (a universal property of any system that depends on feedback loops; imperfection leads to instability and raises the risk of an uncontrollable future disaster). Quantitative easing, the strategy of buying long-term government bonds to put money into circulation, is a glaring example of how inept economists are in dealing with feedback or even knowing what feedback they need. If they had proper STEM education, they would instinctively know how flawed and intellectually bankrupt their thinking is. They need to confront a fundamentally new set of problems, including "how to deal with increasing technological productivity, decreasing real demand for labor, and the effective management of care work, without also destroying the Earth."[37]

The present monetary system drives wealth creating people to passively accumulate their riches because their options to do something useful within the law is very limited. And there are natural psychological limits to ostentatious spending to make others envious. Therefore, wealth lies dormant and becomes unproductive. This situation is now aggravated by rapid advances in AI. The middle class is on its way to becoming poor and destitute. The filthy rich have no knowledge of how to make the unproductive, rote educated poor skilled enough to lead a satisfying, drudgery-free, productive life. The rich can give temporary alms but they cannot secure the future of billions of people who made it to the middle class since the Industrial Revolution but are now sliding down a slippery slope to poverty. Those incapable of exercising genuine intelligence face a dark future. The economic universe we have walked into is new and hence unexplored. It is also mythical.

---

discipline of economics has yet to get over its childish passion for mathematics and for purely theoretical and often highly ideological speculation, at the expense of historical research and collaboration with the other social sciences. … This obsession with mathematics is an easy way of acquiring the appearance of scientificity without having to answer the far more complex questions posed by the world we live in." He adds, "[In France] economists are not highly respected in the academic and intellectual world or by political and financial elites. … [T]hey must set aside their contempt for other disciplines and their absurd claim to greater scientific legitimacy, despite the fact that they know almost nothing about anything.".

[36] Graeber (2019). See also: Werner (2018).
[37] Graeber (2019).

1.8 Money is Abstract

The 20 February 2020 issue of the Economist notes in amazement that in the past 12 months, the shares of the five biggest American tech firms have risen by 52%.

> The increase in the firms' combined value, of almost $2trn, is hard to get your head round: it is roughly equivalent to Germany's entire stockmarket. Four of the five—Alphabet, Amazon, Apple and Microsoft—are each now worth over $1trn. (Facebook is worth a mere $620bn.) For all the talk of a techlash, fund managers in Boston, London and Singapore have shrugged and moved on. Their calculus is that nothing can stop these firms, which are destined to earn untold riches.[38]

The boom is bound to come to a bust and will cause mayhem because it is fueled by greed. On the one hand consumers rail against breach of privacy, on the other they greedily lap up free stuff provided in exchange for voluntarily surrendered personal data and providing free residence to cookies on their computers! So far, the fines and penalties imposed by regulators to date amount to less than 1% of the big five's market value. This is nascent AI pitted against human intelligence. Greed trumps security.

The other aspect of human nature is intrinsic apathy in the relationship between education and men without work. As Nicholas Eberstadt Notes:

> America today is in the grip of a gradually building crisis that, despite its manifest importance, somehow managed to remain more or less invisible for decades .... That crisis is the collapse of work for adult men, and the retreat from the world of work of growing numbers of men of conventional working age.
>
> According to the latest monthly jobs report from the Bureau of Labor Statistics, "work rates" for American men in October 2019 stood very close to their 1939 levels, as reported in the 1940 US Census. Despite some improvement since the end of the Great Recession, Great Depression-style work rates are still characteristic today for the American male, both for those of "prime working age" (defined as ages 25 to 54) and for the broader 20 to 64 group.[39]

This is true not just for America but for all nations. The march of AI is such that the remedy to the "flight from work" problem is not expanding and augmenting rote education and conventional skill building—it will not solve the problem because lack of education did not cause the problem—especially in countries which provide social security and welfare programs to smoothen tough times as a temporary measure. In the same countries, capital distribution is uneven, often from birth due to inheritance laws.

As competition against AI becomes tougher, the natural response of people is to become addicted to social security and welfare programs. Such programs are highly fragile and unsustainable and will collapse in the grip of addiction. The labor market throughout the world has already entered a harsh phase where people of average "intelligence" cannot naturally cope. Economic growth has become disconnected with employment opportunities in the labor market. As the world's current population of 8 billion plus expands to reach almost 11 billion by 2100,[40] and the rising concentration of wealth in the hands of the ridiculously few ("The richest 1% in the

---

[38] Economist (2020).
[39] Eberstadt (2020).
[40] Cilluffo and Ruiz (2019).

world have more than double the wealth of 6.9 billion people"[41]) only means that economic disparity is climbing. There is no hope for the poor.

In disruptive times, we turn to the intelligentsia, while the rest show apathy, freeze into inaction, or fan further disruption. If AI induces speciation of *Homo sapiens* it will force the new species to reexamine the present division between capital and labor in generating income and property, the inequalities induced by the progressive income tax, and the relevance of the welfare state vis-à-vis the duties and responsibilities of the poor to the state. If, as is clear, capitalism mechanically produces grievous inequalities through man-made laws of inherited wealth, should inheritance laws be changed? Do state-funded welfare programs encourage intellectual laziness and encourage apathy? How can AI approach such highly emotional issues when the obvious unemotional solution is to eliminate the cause if possible else eliminate the recipient when funds dry? How does one force *Homo sapiens* to act intelligently, diligently, and eschew malice, greed, and envy? Should religions be taxed for spreading ignorance as tobacco is taxed to discourage smoking?

1.9 Intelligence Versus Artificial Intelligence

> A good idea has a way of becoming simpler and solving problems other than that for which it was intended. – Robert Tarzan

1.9.1 The Path from Data to Wisdom

Creating a path from data to wisdom belongs to the real or imaginary world you want to study or create. It begins with making random observations in that world till something catches your attention, arouses curiosity, and sets you speculating about patterns that may connect your attentive observations (data) into a network. You may think of it as a society of observations and the hierarchical relationships they may have with each other so that a smooth flow of one observation leading to another begins to happen intermittently in a back and forth, hit and trial mental process. That is you make conjectures in your mind and put them on trial by predicting observations not yet made to see if they pass the test. This is the evolutionary process of survival of the fittest mind in the universe of ideas. Your next step is to verify the robustness of your conjectures by venturing to predict observations or behavior patterns in your chosen world that you have not yet found. This is where your innate intelligence to compress observations kicks in. Compression turns observation into useful information. Very few people are gifted with the ability to do this without giving up or losing interest or concluding fairly quickly that he or she and Einstein live in different universes or belong to different species. If this happens to you, you

---

[41] Whiting (2020).

will not survive in the AI-driven world you are stepping into. You will become a gig worker. Yes, our progressing world is brutal.

If you are the curious and curiouser type with the ambition to join the curiousest, there is hope. The cream of the AI (artificial intelligence) fraternity will welcome you with open arms but only after putting you through a grueling test of your ability to see abstract patterns and isomorphically ascribing contextual meaning to them in some real or imaginary world. Skill to detect isomorphism to an extent can be acquired by training under a master, but such masters are rare and they are very choosy as to who they agree to mentor. Much of what you learn and create in your early stages of training even with mentoring can be programmed into a computer. So you will still become a gig worker unless you sharpen your skills in detecting isomorphisms. (The word isomorphism applies when two complex structures can be mapped onto each other in such a way that to each part of one structure there is a corresponding part in the other structure, where *corresponding* means that the two parts play similar roles in their respective structures. In this sense, it is an information-preserving mapping. Since isomorphism may come in various forms, it is very likely that one may miss it when encountered. Physics and mathematics are connected through isomorphisms. Any axiomatized knowledge discipline can be computerized and cast in AI.) Inability to compete against an AI-powered computer will mean job insecurity. Very soon the world will comprise the rich by virtue of their intelligence or highly honed skills in sports, entertainment, and healthcare. The rest will be gig workers with rusted brains. Hoping to stay rich moneywise by inheritance will be a dream and remain a dream. A remote possible exception will be winning a string of lotteries.

The natural intelligence we are born with is fairly rudimentary. It is generally capable of noticing repetitive spatial–temporal arrangements of objects, actions, symbols, ideas, features, attributions, etc., memorizing and recalling them from memory, and effectively mimicking them in standardized, menu-driven conformations that respect peer approval. We assiduously develop a herd mentality to find safety and livelihood in a community where the rote educated lead. The geniuses must usually fend for themselves till chance favors them and puts them under a limelight. With artificial intelligence on the rise, and COVID-19 may yet decimate the *Homo sapiens*, the world will be thrown into a state of chaos. The Darwinian law of the survival of the fittest will ensure the survival of the intelligent and the healthy. Climate change will eventually settle to a new thermodynamic equilibrium with a balanced sequence of seasons, and Nature will prepare to speciate the *Homo sapiens* from a diminished population.

AI as presented to the public now by the experts is as naïve as COVID-19 has been projected by the "experts" to the public: namely, epochal problems are solvable by believable extensions of past ways of problem-solving. To me it appears that COVID-19 is but a trigger for cataclysmic events lurking in the future where climate change will inflict on the environment such changes as to "clean the Augean stables" and AI will enforce speciation of *Homo sapiens* with a brand new neural system. COVID-19 will be the atom bomb that triggers the hydrogen bomb that comprises climate change and AI. Outrageous thoughts! Did we ever imagine that from 17 December 1903 when the Wright brothers (bicycle mechanics) first flew a fragile heavier-than-air flying machine they designed and flew for 59 sec and traveled 852 feet, to Neil Armstrong setting foot on the moon on 24 July 1969 in a mere 65 years was possible? The Turing-complete Electronic Numerical Integrator and Computer (ENIAC), the first programmable, electronic, general-purpose digital computer was completed and put to work in 1945. Is it then so difficult to imagine that AI will surpass the average *Homo sapiens* in intelligence by leaps and bounds pretty soon?

1.9.2 Imitation Versus Innovation

Immature intelligence imitates, mature intelligence innovates. The gap is enormous. The latter is rare. Creating AI is possible only by the intelligent. Without a thorough understanding of the physical sciences and mathematics it is no longer possible for humans to deal with AI machines. AI is advancing just too rapidly. This means that most humans will not be able to deal with superintelligent AI machines, least of all who already find themselves unemployable precisely because they cannot understand science. As Lionel Trilling has noted:

> Physical science in our day lies beyond the intellectual grasp of most men ... This exclusion of most of us from the mode of thought which is habitually said to be the characteristic achievement of the modern age is bound to be experienced as a wound given to our intellectual self-esteem. About this humiliation we all agree to be silent; but can we doubt that it has its consequences, that it introduces into the life of mind a significant element of dubiety and alienation which must be taken into account in any estimate that is made of the present fortunes of mind?[42]

At a psycho-emotional level, indoctrinated irrational ideas and concepts are usually so powerful as to generally bulldoze rational objections by *Homo sapiens* aside. It is only recently that the marvels produced by science and technology have dramatically changed and improved our lifestyle and boosted science in society. On the downside, advancing AI has finally made the average man realize how little his intelligence counts for; he cannot even compete against a machine for a job! The survival of the *Homo sapiens* is now at stake; we are being pushed toward speciation. The new species must exceed our intelligence if they are to survive. Irrational thinking is not intelligent thinking. Quality control for AI machines requires that they function rationally as determined by the axiomatic system they are built on. This means AI developers must know the real meaning of "correspondence" and "predictiveness" as an integral part of intelligent activity. AI must be embedded in a scientific matrix. Construction of a theory is not a mere extrapolation from observations. Theory of relativity and quantum mechanics are resounding examples of that. Such powerful thinking requires a creative leap to be made in abstract mathematics. At each level of understanding, the corresponding theory usually requires counterintuitive and unfamiliar thought processes. A glimpse is provided below.

---

[42] As quoted in Stern (1982).

1.9 Intelligence Versus Artificial Intelligence

Newton thought light consisted of particles, but the idea went out of fashion once Young's experiments showed that it had wave behavior and hence was a wave.[43] In those days physicists believed a thing could be either matter or wave, not both. In 1905, when Einstein explained the photoelectric effect he treated light as a particle (and thereby made a seminal contribution to quantum mechanics) and argued that light was both a particle and a wave. Louis de Broglie then argued that matter particles known as electrons should also have wave behavior. It troubled physicists but eventually it too was experimentally verified. The existence of aether had been postulated for electromagnetic waves to travel through but that too was demolished by Einstein who said that there was no need for it at all and the aether concept too went out of fashion (Occam's razor[44] in action). Current fashions in AI too suffer from dubious ideas, e.g., the ballyhooed neural networks (aka deep learning).[45] Such ideas work only in very narrow areas of knowledge and are easily hackable. AI faces a deep question: "What is intelligence, what does it do?".

To advance AI, inter alia, one needs the ability to hypothesize at various cognitive levels what one observes and state it in a nutshell. It is about one's ability to fathom large-scale structures amid seeming chaos. It is not about research funding; it is all about thinking and a desire to search for nuggets of knowledge that can be assembled into an axiomatic system. Within that system can be found an infinite variety of theorems waiting to be interpreted and connected with the real universe. The theorems provide a treasure trove of knowledge about the universe. Religions singularly fail to provide coherent knowledge; when logic ferrets out paradoxes in that knowledge, miracles are let loose as Band-Aid fixes.

AI presently aims to capture only an objective world whose existence is independent of consciousness. The world we perceive resides in our mind and our perception is the mind's interpretation of our felt sensations. If AI is to mimic human responses based on biological sensors, then it appears we must include biological elements and biochemistry into it. Our malleable minds are in a blank state at birth. In its lifetime, it acquires perceptions through interactions with the environment and introspection. These are saved, curated, and discarded in a self-adaptive manner. Most perceptions are biased by hearsay, indoctrination, nebulously formed but evolving concepts, and personal discoveries that generally assume "local realism" and causality (i.e., the absence of randomness or as Albert Einstein would say, "God does not play dice") in Nature.[46]

---

[43] Young (1802). See also: Rodgers (2002).

[44] Occam's razor is the problem-solving principle: "Entities should not be multiplied without necessity." It is also known as the "law of parsimony".

[45] See, e.g., Hardesty (2017).

[46] Bera (2020). We now know that God does play dice! Quantum mechanics explains how. Physicists readily trust quantum mechanics even without understanding it because it works remarkably well! In fact, it admonishes us to, as David Mermin once quipped, "shut up and compute!" Realism and causality are two distinct ideas; strict determinism implies that after the initial act of creation, God had no further role to play. Realism appeals to common sense and it works only up to a point.

Inexplicably, almost all humans are averse to acting intelligently. They docilely conform, at a basic level, to accepted behavior patterns, belief systems, and social customs to remain socially acceptable with minimal friction and are content with a few opportunities to show off. Had it been otherwise, people would have flocked to imitate Newton, Maxwell, and Einstein, people who made the universe comprehensible. If use of intelligence was pervasive, religions would not exist because there is no viable correlation between praying and prayers being answered. But intelligence can proliferate if it is embedded in machines or genetically enforced via speciation. Humans thrive when and where there is emotion, accommodation, and appreciation. Impeccable rationality is dry, humorless, and tactless. Any discovery of a paradox in a rational system causes grave concern and prompts a search for its origin; in the presence of emotion it often causes dismissive or sardonic humor. Humanity survives because of the intellectual and managerial capabilities of the super few and the deferential obedience to hierarchies under schemes of division of labor by the rest. *Homo sapiens* have socially adapted themselves to an existence of coerced acquiescence.

In the meantime we can be sure that the sudden emergence of COVID-19 pandemic has acted as a trigger to bring in the phase transition I have been expecting. The role of AI will be transformational, and its advancement will be more rapid than envisaged before.[47] Cutting-edge technologies created by advanced countries will not easily percolate to benefit developing and emerging economies. The obvious barriers will be the huge preexisting disparities related to employability factors, capital intensity, and unintelligent labor as compared to the developed economies.

The average *Homo sapiens* loves to wallow in ignorance, spicy gossip, and self-pity. Their irrational behavior can be measured by their natural tenacity to commit or give tacit support to petty crimes. The world is a nasty place because collectively we make it so. Hatred begins with religious strife. Each religion asserts: "We are good no matter what our beliefs about God are; all others are evil. All evil people will end in Hell."

> The true test of intelligence is not how much we know how to do, but how we behave when we don't know what to do.[48]

To this we add Albert Lehninger's perceptive comment:

> ... living organisms are composed of lifeless molecules ... that conform to all the laws of chemistry but interact with each other in accordance with another set of principles—the molecular logic of the living state.[49]

It is this "molecular logic of the living state" that is yet to be completely understood, and therein may lie our ability to understand emotion, cognition, and intelligence. These then can provide a bridge to another set of principles—the sociological logic of the living community. Humans by nature are social animals. To maintain social harmony, intelligence must compromise with irrational emotions. While we

---

[47] See, e.g., Bera (2019).
[48] Holt (1964).
[49] Lehninger (1975).

have a long way to go before machines can compete with Newton, Maxwell, and Einstein in intelligence, we have come a long enough way to augmenting human intelligence at a high level where speciation of the *Homo sapiens* can be accelerated faster than we ever imagined.

References

BBC (2008) Papal visit scuppered by scholars. BBC, 15 Jan 2008. http://news.bbc.co.uk/2/hi/europe/7188860.stm

Bera RK (2008) USPTO, silly patents, and patent trolls. Curr Sci 95(11):1520–1521. https://www.currentscience.ac.in/Volumes/95/11/1520.pdf

Bera RK (2019) Synthetic biology, artificial intelligence, and quantum computing. In: Nagpal ML (ed) Synthetic biology—new interdisciplinary science. IntechOpen, 13 Jan 2019. https://doi.org/10.5772/intechopen.83434. https://www.intechopen.com/books/synthetic-biology-new-interdisciplinary-science

Bera RK (2020) The amazing world of quantum computing. Springer Nature, Singapore. https://doi.org/10.1007/978-981-15-2471-4

Brake ML (2009) Revolution in science: how Galileo and Darwin changed our world. Palgrave Macmillan, p 163. Original in Koestler A (1958) The sleepwalkers. Macmillan Co, New York

Chancel L, Piketty T, Saez E, Zucman G et al (2021) World inequality report 2022. World Inequality Lab, 2022. https://wir2022.wid.world/www-site/uploads/2022/01/Summary_WorldInequalityReport2022_English.pdf

Cilluffo A, Ruiz NG (2019) World's population is projected to nearly stop growing by the end of the century. Fact Tank, Pew Research Center, 17 June 2019. https://www.pewresearch.org/fact-tank/2019/06/17/worlds-population-is-projected-to-nearly-stop-growing-by-the-end-of-the-century/

Cowell A (1992) After 350 years, Vatican Says Galileo was right: it moves. The New York Times, 31 Oct 1992. https://www.nytimes.com/1992/10/31/world/after-350-years-vatican-says-galileo-was-right-it-moves.html

Drake (1957) Galileo Galilei, "Letters on sunspots (1613)". In: Drake S (ed) Discoveries and opinions of Galileo. Doubleday & Co., Anchor Books; Garden City, pp 126–127. https://archive.org/stream/B-001-001-741/B-001-001-741_djvu.txt

Eberstadt E (2020) Education and men without work. National Affairs, Number 42, Winter 2020. https://www.nationalaffairs.com/publications/detail/education-and-men-without-work

Economist (2020) The big tech firms' shares have been on a tear. The Economist, 20 February 2020. https://www.economist.com/leaders/2020/02/20/how-to-make-sense-of-the-latest-tech-surge

Einstein A (1940) Considerations concerning the fundamentals of theoretical physics. Science 91(2369):487–492. In: Address before the eighth American scientific congress, Washington, D. C., 15 May 1940. https://static1.squarespace.com/static/532a9587e4b085a89f267c62/t/551f70a3e4b085a7c1ba05a7/1428123811506/Einstein+commentary+in+Science+1940+.pdf

Eves H (1972) Mathematical circles squared. Prindle, Weber and Schmidt, Boston, 1971

Graeber D (2019) David Graeber reviews Robert Skidelsky's money and government: the past and future of economics. Yale University Press, in the New York Review of Books essay, 05 Dec 2019. https://www.nybooks.com/articles/2019/12/05/against-economics/

Halsall P (1999) Documents in the case of Galileo: indictment, sentence and abjuration of 1633. July 1998, rev. January 1999. https://hti.osu.edu/sites/default/files/documents_in_the_case_of_galileo.pdf

Hardesty L (2017) Explained: neural networks. MIT News, 14 Apr 2017. http://news.mit.edu/2017/explained-neural-networks-deep-learning-0414

Holt JC (1964) How children fail. Penguin Education, p 163. http://www.schoolofeducators.com/wp-content/uploads/2011/12/HOWCHILDREN-FAIL-JOHN-HOLT.pdf

Israely F (2008) (No title). Time, 15 Jan 2008. https://vaticandiplomacy.wordpress.com/tag/la-sapienza/page/8/

Juan S (2006) What are the most widely practiced religions of the world? The Register, 06 Oct 2006. https://www.theregister.co.uk/2006/10/06/the_odd_body_religion/

Kirsch A (2018) Why Jewish history is so hard to write. The New Yorker, 19 Mar 2018. https://www.newyorker.com/magazine/2018/03/26/why-jewish-history-is-so-hard-to-write

Landauer R (1991) Information is physical. Phys Today 44(5):23. https://doi.org/10.1063/1.881299. http://www.w2agz.com/Library/Limits%20of%20Computation/Landauer%20Article,%20Physics%20Today%2044,%205,%2023%20(1991).pdf

Lehninger AL (1975) Biochemistry, 2nd edn. Worth Publishers, New York

Lynch P (2018) The book of nature is written in the language of mathematics. The Irish Times, 15 Feb 2018. https://www.irishtimes.com/news/science/the-book-of-nature-is-written-in-the-language-of-mathematics-1.3388465

Mariani MS, Ren Z-M, Bascompte J, Tessone CJ (2019) Nestedness in complex networks: observation, emergence, and implications. Phys Rep 813:1–90 https://www.sciencedirect.com/science/article/pii/S037015731930119X

New Scientist (1992) Vatican admits Galileo was right. New Scientist, 7 Nov 1992. https://www.newscientist.com/article/mg13618460-600-vatican-admits-galileo-was-right/

Newton I (1687) Philosophiæ naturalis principia mathematica (1st edn, New Latin). English translation (1729) by Andrew Motte based on the 1726 3rd edition of Philosophiae Naturalis Principia Mathematica. https://en.wikisource.org/wiki/The_Mathematical_Principles_of_Natural_Philosophy_(1729)

PAS (2003) Papal addresses. The Pontifical Academy of Sciences. Scripta Varia 100. Vatican City. http://www.pas.va/content/dam/accademia/pdf/sv100.pdf

Piketty T (2014) Capital in the twenty-first century. Translated by Arthur Gold hammer. Harvard University Press

Plotkin H (1997) Darwin machines and the nature of knowledge. Harvard University Press

Popper K (1934) The logic of scientific discovery. Routledge, 1959. (Originally published in German: Logik der Forschung, Mohr Siebeck, 1934)

Popper K (1963) Conjectures and refutations: the growth of scientific knowledge (Reprinted Ed., 2004). (First published, 1963), Routledge

Renn J (2020) The evolution of knowledge: rethinking science for the Anthropocene. Princeton University Press

Rodgers P (2002) The double-slit experiment. Physics World, 01 Sept 2002. https://physicsworld.com/a/the-double-slit-experiment/ and https://www.physics.umd.edu/courses/Phys401/appeli/EXTRAS/double-slitexperiment.pdf

Rose N (1988) Mathematical maxims and minims. Rome Press Inc., Raleigh NC

Shannon CE (1948) A mathematical theory of communication. Reprinted with corrections from Bell Syst Tech J 27:379–423, 623–656. https://www.cs.ucf.edu/~dcm/Teaching/COP5611-Spring2012/Shannon48-MathTheoryComm.pdf

Stanley A (2000) Honoring a heretic whom Vatican 'Regrets' burning. The New York Times, 18 Feb 2000. https://www.nytimes.com/2000/02/18/world/honoring-a-heretic-whom-vatican-regrets-burning.html

Stern F (1982) Einstein's Germany. In: Goldsmith D, Bartusiak M (eds) E=Einstein: his life, his thought, and his influence on our culture. Sterling Publishing, New York, pp 97–118. The Stern article originally appeared in *Albert Einstein: Historical and Cultural Perspectives*, Holton G, Elkana Y (ed). Princeton University Press

Stewart M (2019) Security vulnerabilities of neural networks. In: Towards data science, 24 Apr 2019. https://towardsdatascience.com/hacking-neural-networks-2b9f461ffe0b?gi=10bf335b230e

References

Stiglitz JE (2020) Solidarity now. Project Syndicate, 28 Feb 2020. https://www.project-syndicate.org/onpoint/abandoning-neoliberalism-reforming-capitalism-globalization-by-joseph-e-stiglitz-2020-02

TCE (1912) Heresy. The Catholic Encyclopedia. New advent. 1912. http://www.newadvent.org/cathen/07256b.htm

Turing AM (1936) On computable numbers, with an application to the *Entscheidungsproblem*. Proc London Math Soc S2–42(1):230–265, 1936–1937. https://www.cs.virginia.edu/~robins/Turing_Paper_1936.pdf Correction at: Turing AM (1938) On computable numbers, with an application to the Entscheidungsproblem. A Correction. S2–43(1):544–546. http://www.turingarchive.org/viewer/?id=466&title=02

USA Today (2006) Stephen Hawking says Pope told him not to study beginning of universe, USA Today, 15 June 2006. http://usatoday30.usatoday.com/life/people/2006-06-15-hawking_x.htm

Werner R (2018) Conversation with Prof. Richard Werner. Moderator: Stefan Grobe. Published on Oct 12, 2018. https://www.youtube.com/watch?v=8FT-zyTX2nE

Whiting K (2020) 5 shocking facts about inequality, according to Oxfam's latest report. World Economic Forum, 20 Jan 2020. https://www.weforum.org/agenda/2020/01/5-shocking-facts-about-inequality-according-to-oxfam-s-latest-report/

Young T (1802) On the theory of light and colours (The 1801 Bakerian Lecture). Philos Trans Roy Soc London 92:12–48

Zax D (2009) Galileo's revolutionary vision helped usher in modern astronomy. Smithsonian Magazine, Aug 2009. https://www.smithsonianmag.com/science-nature/Galileos-Revolutionary-Vision-Helped-Usher-In-Modern-Astronomy-34545274/

Zukav G (1979) Dancing Wu Li Masters. Rider, London

Chapter 2
Scientific Theories Are Intellectual Constructs

Abstract We are experiencing the birth pangs of a new era in the lives of the *Homo sapiens* because of rapid advances in AI, robotics, and automation and the resulting frictional changes they impose on industry dynamics, socio-economic fundamentals, and the meaning of market competition. The central lesson we derive in this book is that a rapid rate of progress or too much connectivity comes at a price one may not always want to pay or even know how to deal with.

Keywords Scientific theories · Artificial intelligence · Knowledge · Survival · Rationalism · Evolution

2.1 The Search for Scientific Theories

Since Galileo Galilei (1564–1642), physicists have adopted mathematics as their lingua franca, and since Isaac Newton (1642–1727), additionally, following the advice of William of Ockham (1285–1347) that "plurality should not be posited without necessity" (famously known as "Ockham's razor"), the axiomatic system to describe Nature. Thus, physicists have mastered the art of relating the physical, observable, and inferable aspects of Nature in the language of mathematics in a lean and mean way and isomorphically attaching meaning to mathematical symbols by creating consistent 'symbol-meaning' pairings to interpret Nature. Because it is abstract, mathematics serves as a universal language and because it has carefully crafted grammar, it lacks ambiguity or ambiguities are detectable—that is, consistent statements in mathematics are provable as either true or false (but never both at the same time) in a prescribed manner. The fact that any axiomatic system can be arithmetized, means that arithmetic serves as a canonical axiomatic system in which all other axiomatic systems can be represented isomorphically.[1]

When we ascribe meaning to an axiomatic system by interpreting it, we must exercise extreme caution. Creating symbol-meaning pairing requires a high level of "intelligence" to ensure consistency in the meaning. A faulty pairing leads to an

---
[1] Hofstadter (1979).

erroneous understanding of the world. It is, therefore, instructive to see how we have fared in refining our understanding of something, say, as fundamental as space and time since the time of Aristotle.

2.1.1 Theories Should Be Lean and Mean

Following the lean and mean prescription of Ockham's razor assiduously, mathematicians parsimoniously choose their symbols that stand for operands and operators, a minimal set of axioms, and rules for symbol manipulation that produce valid new strings of symbols they call theorems. Given appropriate machinery capable of performing one or more operations and the ability to self-start, a universe, in principle, can be created. That primitive machinery is time, and the first operands it created were energy, force, and space, and the new operators it created were motion and rate of change of motion. The essence of time is to ceaselessly sequence change (motion) in the distribution of operands, and the creation, destruction, and modification of operands and operators. Time is the creator of the manifest universe, and this universe is a self-referential (iterative) system capable of creating *inter alia* fractal objects and strange attractors.

2.1.2 How Science Progresses

Science is dominated by brutal reasoning; religion is dominated by emotion and mindless devotion. Science struggles against the fallibility of human understanding; religion promotes unverifiable assertions as absolute truth. Science thrives on doubt, religion on not entertaining it. Science is not in search of truth; it strives to deal with doubts by making conjectures. As Richard Feynman notes:

> [I]t is imperative in science to doubt; it is absolutely necessary, for progress in science, to have uncertainty as a fundamental part of your inner nature. To make progress in understanding we must remain modest and allow that we do not know. Nothing is certain or proved beyond all doubt. You investigate for curiosity, because it is unknown, not because you know the answer. And as you develop more information in the sciences, it is not that you are finding out the truth, but that you are finding out that this or that is more or less likely.[2]

Here is an example of how Einstein viewed the process:

> I am glad that our colleagues are busying themselves with my theory—even if it's with the hope of killing it. – Albert Einstein, in a letter to Erwin Finlay Freundlich, August 7, 1914.[3]

---

[2] Feynman (2005).

[3] As quoted in Aczel (1999) in Goldsmith and Bartusiak (2006), pp. 81–95. The Aczel article is from *God's Equation: Einstein, Relativity and the Expanding Universe*. Thunder's Mouth Press, 1999.

2.1.3 The Web and the Mind

In his book, *Smart World: Breakthrough Creativity and the New Science of Ideas,* Richard Ogle makes the insightful observation:

> [I]n making sense of the world, acting intelligently, and solving problems creatively, we do not rely solely on our mind's internal resources. Instead, we constantly have recourse to a vast array of culturally and socially embodied idea-spaces ... in forms as various as myths, business models, scientific paradigms, social conventions, practices, institutions, and even computer chips [to which] we have progressively offloaded ... for the sake of simplifying the burden on our own minds of rendering the world intelligible. Sometimes the space of ideas thinks for us.[4]

The way we acquire knowledge is iterative and nonlinear—we conjecture and put our conjectures on trial. As the trial progresses, we edit, discard, refine, add to our conjectures in a pseudo-random manner driven by instinct, hunches, inspiration, serendipity, etc. We connect the dots. At every step of linking the dots we consult the axioms (conjectures) and the rules for deriving conclusions (theorems) to ensure that we are within the axiomatic system we have put on trial. The process leads us to understand the universe based solely on our beliefs (axiomatic system) we put our faith in. We can live in the world of Euclid or of Mandelbrot[5] or of another or of our own imagination. The choice of how we connect the dots is ours. To a researcher the game is worth living for. As Karl Popper notes:

> Our aim as scientists is objective truth; more truth, more interesting truth, more intelligible truth. We cannot reasonably aim at certainty. Once we realize that human knowledge is fallible, we realize also that we can never be completely certain that we have not made a mistake.[6]

Evolutionary biology emphatically tells us that God did not create man in His own image, rather man created God in his own image. *Homo sapiens* is not the ultimate goal of evolution; life began as single-cell organism and evolved via speciation from there. Rene Descartes said, "I think, therefore I am". If he had pondered, he would have known "I am, therefore I think" is also true.

---

[4] Ogle (2007). Quote reproduced from https://richardogle.typepad.com/site/smart_world_excepts/.

[5] Benoit Mandelbrot (1924–2010) is the father of fractal geometry and creator of perhaps the most complicated mathematical object yet known to man, the Mandelbrot set. For a quick introduction to Mandelbrot see Matson (2010). "Mandelbrot made the case that fractals could help make sense of everything from the shape of coastlines to the performance of Wall Street.".

[6] Popper (1994), p. 4.

2.1.4 On Building Theories

Here is a profound insight from Gregory Chaitin:

> [A] scientific theory is a computer program that calculates the observations, and that the smaller the program is, the better the theory. If there is no theory, that is to say, no program substantially smaller than the data itself, considering them both to be finite binary strings, then the observations are algorithmically random, theory-less, unstructured, incomprehensible and irreducible.[7]

This is the sum and substance of artificial intelligence. But there is a catch. For very practical reasons, we never begin with an enormous amount of data. We always begin with a small amount of data, make bold and daring hypotheses, and then generalize to see if the unused data falls in line with our hypotheses. Thus begins the unending process of refining and modifying hypotheses and sometimes discarding hypotheses and starting anew. The process is fraught with hidden dangers. For example, how can we know that different people starting with different lots drawn from a huge repository of data will converge to a common theory and not develop along irreconcilable paths which, nevertheless, within the bulk of the data analyzed provide reasonable theories in terms of their predictive power? We have no reasonable evidence that this can never happen for the simple reason that the vastness of the data repository does not guarantee that the data is neither skewed, nor biased, nor blindsided. There is another problem which leads to incredulity. For example, in many mathematical formulas we find the symbol π, an irrational number, standing for the universal ratio of the circumference of the circle to its diameter. For example, in statistics, the Gaussian distribution contains π and the distribution is frequently used to arrive at a simplified, tractable model when working on population trends. This may make one wonder how population trends get mixed up with the radius and the circumference of a circle![8] The connection is deep and not at all obvious because it is at an abstract level that it ties together certain symmetries and repetitive patterns.

Such examples show that mathematical concepts show up in unexpected and surprising ways when abstract concepts are paired with meanings in the real world. It amazes one to see that mathematics does indeed provide uncannily close and accurate description of certain phenomena by such pairings. This enormously useful aspect of mathematics, while delightfully magical, raises doubts about the uniqueness of the physical theories we conjure. Maybe, like the picture of "My wife and my mother-in-law", or the musician and a beautiful woman (see Fig. 2.1) mathematics and physics are inseparably woven into each other. Maybe mathematics and physics are two aspects of the same thing. Perhaps we need a few more hints in addition to the insights of Leopold Kronecker (1823–1891) "God made the integers, all the rest is the work of man"[9]; Gregory Chaitin (1947–) "a scientific theory

---

[7] Chaitin (2003).

[8] The two examples are provided in Wigner (1960).

[9] He said it in German: "Die ganzen Zahlen hat der liebe Gott gemacht, alles andere ist Menschenwerk."

Agree to disagree (Nonuniqueness of interpretations)
A logician says, "I am a liar!" Is he a liar?

Someone says, "This is a picture of my son, a musician." His friend says, "You are joking. This is your daughter!"

An artist says, "This is a portrait of my wife." His friend says, "This is your mother-in-law!"

Who is correct? Possibly both!

Fig. 2.1 Agree to disagree. *Source of figure* (Left) Sara Nader by Roger Shepard. https://upload.wikimedia.org/wikipedia/commons/8/85/MooneyFaces.jpg. This file is licensed under the Creative Commons Attribution-Share Alike 3.0 Unported license. (Right) My wife and my mother-in-law, by the cartoonist W. E. Hill, 1915 (adapted from a picture going back at least to a 1888 German postcard). https://commons.wikimedia.org/wiki/File:Youngoldwoman.jpg (public domain)

is a computer program"[10]; Max Tegmark (1967–) "Our reality isn't just described by mathematics—it is mathematics"[11]; Rolf Landauer (1927–1999) "Information is physical"[12]; Benoit Mandelbrot (1924–2010) "Clouds are not spheres, mountains are not cones, coastlines are not circles, bark is not smooth, nor does lightning travel in a straight line"[13]; and Claude Shannon (1916–2001) "Information is the resolution of uncertainty".[14] Indeed, in 1987, Shannon said, "I visualize a time when we will be to robots what dogs are to humans, and I'm rooting for the machines."[15] Shannon, in addition to his famous theory of communication, published a classic paper *A Symbolic Analysis of Relay and Switching Circuits*, which showed the identity between the two "truth values" of symbolic logic and the binary values 1 and 0 of electronic circuits. He showed how a "logic machine" could be built using switching circuits corresponding to the propositions of Boolean algebra.[16]

---

[10] Chaitin (2003).
[11] Tegmark (2014).
[12] Landauer (1991).
[13] Mandelbrot (1982), p. 1.
[14] One may conclude this from Shannon (1948).
[15] Source of quote: Simske (2019), Chap. 12, p. 287.
[16] Shannon (1937, 1938). See also: Claude Shannon. http://www.nyu.edu/pages/linguistics/courses/v610003/shan.html.

2.2 An Obsession with Symmetry

As in art, so in science and mathematics. The notion of symmetry is intimately entwined with our perception of beauty. It is why we seek unity in diversity. It led Eugene P. Wigner (1902–1995) in 1960 to make, in amazement, the often quoted statement, *"The power and unreasonable effectiveness of mathematics in the natural sciences."*[17] It led another famous physicist, Richard P. Feynman, to comment,

> When learning about the laws of physics you find that there is a large number of complicated and detailed laws, laws of gravitation, of electricity and magnetism, nuclear interactions, and so on. But across the variety of these detailed laws there sweep great general principles which all the laws seem to follow. Examples of these are the principles of conservation, certain qualities of symmetry ...[18]

We shall see an amazing example of this in Sect. 2.2.4 which describes certain fundamental symmetries in Nature related to conservation of momentum and energy. There is awe and enchantment. Today, physicists search for as many symmetries in Nature as they can get. It is a guiding star in their speculations. Lack of symmetry, when intuitively expected, brings a troubled look on their faces. When we identify an object, in spite of distortions, additions, and deletions, we do it on the basis of some invariant qualities of the object, that is, we identify them on the basis of symmetry. In the twentieth century, physicists have, one may say, been obsessed with invariances that govern Nature, seeing in them both beauty of symmetry and deep simplicity. One naturally wishes to select the conserved quantities as the physical quantities with which to describe Nature. Another way of looking at the laws of Nature is that they are laws of prohibition, that is, they prohibit any phenomenon that would change the conserved quantity.

Historically, theorists in natural sciences, and particularly in physics, have preferred those theories, which stand out as having logical elegance and beauty. Beauty is subjective and has no rigorous value, but the first quality is objective to the extent that it can be used in practice. In addition, they prefer simplicity, which is again subjective. To study Nature, look for simplicity, look for beauty, and look for symmetry.

2.2.1 What is Symmetry in Mathematics?

A thing is symmetrical if there is something we can do to it so that after it is done, it still looks the same. For example, if you rotate a perfect and rigid sphere or even when such a sphere is rotating about any axis through its center, you will not know that it is rotating by seeing it with your unaided eyes and without interfering with its motion. To mathematicians and physicists: *symmetry is invariance under transformations.*

---

[17] Wigner (1960).
[18] Feynman (1965).

2.2 An Obsession with Symmetry

That is, the end state looks the same after the transformation T as the start state. It means $f(x) = f(y = T(x)) = f(y)$. The transformation only changes the symbols defining the function f but not its form. Here is an example:

Given a transformation T and an expression E, say,

$$T : x = x' \cos q - y' \sin q, \, y = x' \sin q + y' \cos q \text{ and } E : ax + by + c,$$

then on substituting for x and y using T in E we get

$$a'x' + b'y' + c', \quad \text{where}$$

$$a' = a \cos q + b \sin q, \, b' = -a \sin q + b \cos q.$$

It therefore follows that

$$a'^2 + b'^2 = a^2 + b^2.$$

Accordingly, $a^2 + b^2$ is called an *invariant of E* under every transformation of the type T.

Generally, if a mathematical structure (expression, equation, etc.), $S(V)$, undergoes a transformation, T, where its set of variables, $V = \{v_1, v_2, \ldots, v_n\}$ is replaced with another set of variables $V' = \{v'_1, v'_2, \ldots, v'_n\}$ in terms of the elements of V and produces $S'(V')$, where $V' = T(V)$, and if certain features between S and S' remain the same except that unprimed variables are substituted by primed variables, then those features are invariants of S under every transformation of the type T. As a general rule, whenever mathematical Invariants exist, they usually provide deep insight both in mathematics and the physics they describe. The reader should pay great attention to invariants whenever and wherever they appear in any branch of knowledge. They are absolutely crucial in the design of AI systems.

2.2.2 Inertial Reference Frame

An inertial reference frame is a coordinate system which moves with constant velocity, i.e., it maintains its speed and direction. Since the reference frame is not accelerating therefore any object with mass fixed to it will not experience a force. Its profound implication is that if you are that object you will not know if you are standing still in the universe or moving in it. You will perceive only relative motion between yourself and the universe. Within this limited perception you can perceive and measure your displacement and relative velocity with respect to another object. Absolute measurements of location, direction, speed, velocity, time are out; relative measurements are in with respect to references we choose. That means that each of us can choose and personalize our references as often as we wish and have a viewpoint of our own with respect to it. Measurements made of an object or event can always be put in correspondence with like measurements of the same object or event made using a different inertial frame with mathematical precision using the Galilean transformation. This was an early form of symmetry discovered in physics, and it became an intuitive part of our understanding of Nature. This symmetry means that the intrinsic laws of Nature (e.g., Newton's laws of motion) act without fear or favor equally at every point in space and time in the universe. As we shall see in Sect. 2.2.4, Newton's Laws of Motion are derivable consequences of it. Simple symmetries as perceived by *Homo sapiens* in Nature have deep derivative consequences in physics.

2.2.3 Maxwell's Equations of Electromagnetism

Here is an even more enigmatic example. An unusual symmetry appears in Maxwell's equations of electromagnetism conjectured by James Clerk Maxwell in 1865. His equations are:

$$\nabla \cdot E = \frac{\rho}{\varepsilon_0}, \quad \nabla \times \mathbf{E} = -\frac{\partial \mathbf{B}}{\partial t}$$

$$\nabla \cdot \mathbf{B} = 0, \quad c^2 \nabla \times \mathbf{B} = \frac{\mathbf{j}}{\varepsilon_0} + \frac{\partial \mathbf{E}}{\partial t}$$

$$\mathbf{F} = q\mathbf{E} + q\mathbf{v} \times \mathbf{B}$$

Here $\mathbf{E}$ = electric field vector, $\mathbf{B}$ = magnetic field vector, ρ = electric charge density (charge per unit volume), $\mathbf{j} = \rho \mathbf{v}$ = electric current density (charge flowing through per unit area per unit time), $\mathbf{v}$ = average drift velocity of the charges, c = speed of light, t = time, ε_0 = a constant, $\mathbf{F}$ is the force field vector, and q is the charge. The equations describe the time evolution of the electric and the magnetic fields where the electric charge density and the electric current density are considered as given. The constants ε_0 and c can be determined from experiments.

Maxwell's equations unified electricity and magnetism. It was a remarkable breakthrough in classical physics. In a static world electricity and magnetism are separate phenomena, and in a dynamic world they are not. Yet, for many years Maxwell's equations were met with skepticism since they were not invariant with respect to the Galilean transformation. It was believed at the time that any law of Nature should be so invariant just as Newton's Laws of motion were.[19] Yet, experiment after experiment confirmed Maxwell's equations. Some 20 years later, Hendrik Antoon Lorentz (1853–1928) noticed that the equations are invariant under the transformation now known by his name:

---

[19] Dyson (2007).

2.2 An Obsession with Symmetry

$$x' = \frac{x - ut}{\sqrt{1 - u^2/c^2}}, \; y' = y, \; z' = z, \; t' = \frac{t - ux/c^2}{\sqrt{1 - u^2/c^2}},$$

where for simplicity, shifts in origin have been ignored, and the relative velocity u is assumed to be in the x-direction. Under the Lorentz transformation, time and distance between events may differ among inertial reference frames; however, the Lorentz scalar distance s^2 between two events is the same in all inertial reference frames, i.e.,

$$s^2 = (x_2 - x_1)^2 + (y_2 - y_1)^2 + (z_2 - z_1)^2 - c^2(t_2 - t_1)^2.$$

$c = \infty$ gives the Galilean transformation. In the Lorentz reference frame, c is only coincidently the speed of light; fundamentally it is a property of space–time, a conversion factor between time units and length units.

The constant c^2 is the same no matter what we choose for our unit of charge. Experiments show that c^2 is the square of the velocity of propagation of electromagnetic influences. Using available data, Maxwell obtained $c = 310,740,000$ m/second. It was not obvious that the coefficient c in Maxwell's equations was also the speed of light propagation, but Maxwell did notice the mysterious coincidence. In an 1864 lecture titled *A Dynamical Theory of the Electro-Magnetic Field*, before the Royal Society of London, Maxwell asserted:

> We have strong reason to conclude that light itself—including radiant heat and other radiation, if any—is an electromagnetic disturbance in the form of waves propagated through the electro-magnetic field according to electro-magnetic laws.[20]

Numerous experiments have proved Maxwell correct even at the level of quantum mechanics. Finally, Albert Einstein would use the Lorentz transformation and the fact that the constant c^2 is the same no matter what we choose for our unit of charge, to show that space and time are indeed related. Thus was born Einstein's special theory of relativity.[21]

While paying a tribute to Michael Faraday in his book, Maxwell wrote:

> When I had translated what I considered to be Faraday's ideas into a mathematical form, I found that in general the results of the two methods coincided, so that the same phenomena were accounted for, and the same laws of action deduced by both methods, but that Faraday's methods resembled those in which we begin with the whole and arrive at the parts by analysis, while the ordinary mathematical methods were founding on the principle of beginning with the parts and building up the whole by synthesis.[22]

It is even more interesting to learn how Maxwell approached his study of electromagnetism. He explains:

> [B]efore I began the study of electricity I resolved to read no mathematics on the subject till I had first read through Faraday's Experimental Researches in Electricity. I was aware

---

[20] Britannica (n.d.).
[21] Einstein (1905).
[22] Maxwell (1873).

that there was supposed to be a difference between Faraday's way of conceiving phenomena and that of the mathematicians, so that neither he nor they were satisfied with each other's language. I had also the conviction that this discrepancy did not arise from either party being wrong. I was first convinced of this by Sir William Thomson, to whose advice and assistance, as well as to his published papers, I owe most of what I have learned on the subject.

As I proceeded with the study of Faraday, I perceived that his method of conceiving the phenomena was also a mathematical one, though not exhibited in the conventional form of mathematical symbols. I also found that these methods were capable of being expressed in the ordinary mathematical forms, and thus compared with those of the professed mathematicians.[23]

AI researchers can learn an extremely valuable lesson here. We generally approach a new field of knowledge first by looking at the whole and then gravitate to mathematical-computational methods and arrive at the parts by analysis from which we synthesize the whole. The building of AI systems needs both approaches when the objective is to find a rational hypothesis that can bind unexplained data. The big picture captures our attention first, the interaction among the smaller elements usually takes time and requires greater attention to details. Once we begin to understand the parts, we intuitively begin to surmise many other bigger pictures that may also be built. I believe, we will understand neural networks in AI much better once we understand information flow between neurons in the living brain.

2.2.4 Nöther's Theorem

Amalie Emmy Nöther (1882–1935), an associate of David Hilbert (1862–1943), working along with Felix Klein (1849–1925), and Albert Einstein (1879–1955) on the mathematical foundations of physics and relativity proved a beautiful theorem in 1918:

If the first integral of a function of generalized coordinates and first derivatives is invariant under an infinitesimal transformation, then the first integral of the related Euler-Lagrange equation is a constant.[24]

In simpler language: for every observable symmetry in Nature there is a corresponding entity that is conserved. And for every conservation law there is a corresponding symmetry. It holds for all physical laws based upon the action principle.

The action is an integral quantity that is used to determine the evolution of a physical system between two defined states, say, $x(t_1)$ and $x(t_2)$, using the calculus of variations. The requirement that the action integral be stationary under small perturbations of the evolution is equivalent to a set of differential equations (called the Euler–Lagrange equations). If we assume that the function L (the integrand of the action integral) depends only on the coordinate $x(t)$ and its time derivative

---

[23] Maxwell (1873).
[24] Nöther (1918).

2.2 An Obsession with Symmetry

$\dot{x} = dx(t)/dt$ and does not depend on time explicitly then the action integral can be written as

$$S = \int_{t_1}^{t_2} L(x, \dot{x}) dt.$$

The Euler–Lagrange equation is then (from variational calculus)

$$\frac{\partial L}{\partial x} - \frac{d}{dt} \frac{\partial L}{\partial \dot{x}} = 0.$$

In mechanics, $L = K - V$, where K is the total kinetic energy and V is the total potential energy of the system. Nöther's theorem (completed in 1915 but published in 1918) applies only to continuous and smooth symmetries over physical space. Lagrangian mechanics is a reformulation of classical mechanics, introduced by Joseph-Louis Lagrange in 1788. The following results follow from the Nöther theorem:

(1) *The law of conservation of momentum is a consequence of the homogeneity of space.* Viewed differently, it is a consequence of the invariance of the laws of physics under translations in space. In layman terms: there are no absolute positions; what matters is not where an object is in absolute terms, but where it is relative to other objects.
(2) *The law of conservation of angular momentum is a consequence of the isotropy of space.* That is, it is a consequence of the invariance of the laws of physics under rotations in space.
(3) *The law of conservation of energy is a consequence of the homogeneity of time.* That is, it is a consequence of the invariance of the laws of physics under translations in time.

What does it all imply? Inter alia, a mathematical law of physics involving the x coordinate of two objects x_1 and x_2 may contain only their differences but not their absolute values. The law can have the form $f(x_2 - x_1) = 0$ but not $f(x_1) = 0$ or $f(x_2) = 0$. Or, even for a single object, it may contain the time derivative of its x coordinate since $d(x + c)/dt$ is the same as dx/dt for any constant c. Such equations will remain invariant under a translation. There is one common characteristic to these transformations under which physical laws remain unchanged. Namely that all are *global*: the value that characterizes the transformation is the same everywhere in all of space-time.

The three-dimensionality of space predetermines the vector nature of momentum and angular momentum, and the laws of conservation of momentum and angular momentum are vector laws. The one-dimensionality of time predetermines the scalar nature of energy and the corresponding conservation law. The relationship of conservation laws to space–time symmetry means that the passage of time or a translation and a rotation in space cannot cause a change in the physical state of the system.

The laws of momentum, angular momentum, and energy are used both in classical mechanics and in quantum mechanics. So far no experiment has shown that the laws of nature may be variant under translation in time, and under translation and rotation in space. It is interesting that Euclid's fourth axiom (all right angles are congruent), in effect, asserts the isotropy and homogeneity of space, so that a figure in one place could have the same (i.e., congruent) geometrical shape as a figure in some other place.

The law of conservation of energy had been used in mechanics before Galileo. In fact, in the late fifteenth century the great Leonardo da Vinci postulated the impossibility of perpetuum mobile. In his book *On True and False Science* he wrote: "Oh, seekers of perpetual motion, how many empty projects you have created in those searches." The laws of conservation of linear momentum and angular momentum were formulated later, in the 17–eighteenth centuries. It was not, however, till the beginning of the twentieth century that the laws assumed prominence. The attitude to them changed radically only after it was discovered that *these laws were related to principles of invariance*. Once this relation had been revealed, it became clear that conservation laws are predominant among other laws of nature.

The analog of Nöther's theorem in quantum field theory is the Ward-Takahashi identities,[25] which lead to results such as the conservation of electric charge from the invariance with respect to the gauge invariance[26] of the scalar electric potential and vector magnetic potential. Nöther's theorem is also valid in quantum field theory.

Nöther's theorem allows one to gain powerful insights into the laws of Nature when one analyzes transformations that leave the form of the laws invariant. The theorem is deeply tied to quantum mechanics as it identifies physical variable pairs that are related by the Heisenberg uncertainty principle (such as position—momentum, and time—energy) using only the principles of classical mechanics. The formal statement of the theorem derives from the condition of invariance alone, an expression for the current associated with a conserved physical quantity. The conserved quantity is called the Nöther charge and the current the Nöther current, which is defined by a divergence free vector field.

This is a profound example of intelligence in action. Insight is about finding deep relationships between variables by drawing inspiration from the environment (in this case an amazing group of talented scientists she was associated with). Machines interacting among themselves have no reason to be intelligent. We want machines to be intelligent in a human environment. AI machines can deliver superintelligence if they interact with intelligent humans; rote educated humans do not fit the bill. The most important aspect of intelligence is the ability to ask questions driven by curiosity and then striving to answer them. Curiosity can be programmed, e.g., by algorithmically or randomly generating symbol strings and presenting them to an

---

[25] Danos (1997). "The gap in the mathematical derivation of Nöther's theorem, and also of the Ward-Takahashi identities, caused by performing variation before quantization is closed by introduction of variational calculus for operator fields. It is demonstrated that both Nöther's theorem and the Ward-Takahashi identities retain full validity in quantum field theory.".

[26] For a brief description of gauge invariance, visit Toth (2003).

2.2.5 Symmetry in Quantum Mechanics

Interestingly, quantum mechanics makes it possible to show the fundamental nature of conserved quantities and relate them to corresponding types of symmetry in the laws of physics. For each symmetry in physics there is a corresponding conservation law in quantum mechanics: energy, momentum, charge, etc. Even the Schrödinger equation of quantum mechanics has a symmetry. This equation is unchanged if the phase of the wave function ψ is shifted by an arbitrary constant φ. That is, you will not be able to detect any difference between a system described by ψ and another by $\psi e^{i\varphi}$. In fact, this symmetry is related to the *conservation of electrical charge*. Symmetries are among the most beautiful and profound things in physics.

In classical physics, energy E, momentum **p**, and angular momentum **M** appear as functions of the velocity **v** and coordinates r of a body of mass m as:

$$E = \frac{mv^2}{2}, \mathbf{p} = m\mathbf{v}, \mathbf{M} = (\mathbf{r} \times m\mathbf{v}),$$

from which it follows that

$$E = \frac{p^2}{2m}, \mathbf{M} = (r \times \mathbf{p}).$$

However, when we go into the microworld of quantum mechanics, the very concept of the velocity of an object becomes unsuitable and these formulas become pointless. At the same time the conserved quantities (E, **p**, **M**) retain their meaning both in classical and quantum mechanics. But, very importantly, in quantum mechanics (E, **p**, **M**) are, generally speaking, not expressible in terms of each other. The second expression, namely

$$E = \frac{p^2}{2m}, \mathbf{M} = (r \times \mathbf{p}),$$

does not hold in the microworld, since a micro-object has no states in which the values of momentum and coordinates can be specified simultaneously. This follows from Heisenberg's uncertainty principle. The first expression, namely

$$E = \frac{mv^2}{2}, \mathbf{p} = m\mathbf{v}, \mathbf{M} = (\mathbf{r} \times m\mathbf{v}),$$

is valid only for the free motion of a micro-object. For a bound micro-object (say, an atomic electron) the energy is quantized, with the result that for each energy level we cannot indicate a definite value of momentum.

2.2.6 Matter and Anti-matter

In 1928 Paul Dirac (1902–1984) wrote down an equation, which combined quantum theory with special relativity to describe the behavior of an electron. His equation had two possible solutions, one for an electron with positive energy, and another for an electron with *negative* energy. Rather than abandon his equation, he boldly interpreted the electron with the negative energy to imply that for every electron there must be an anti-electron, identical in every way but with a positive electric charge. To postulate the existence of matter-anti-matter was an amazing counterintuitive belief in symmetry in the laws of Nature.

It turns out that without antiparticles, the equations of theoretical physics describing the various types of elementary particles would not be invariant under the Lorentz transformation. So the electron has an anti-electron (positron), the proton has an anti-proton, the photon is its own anti-photon, etc. The positron was later experimentally discovered by Carl David Anderson (1905–1991) at Caltech in 1932. Paul Dirac was awarded the 1933 Nobel Prize in physics for predicting the existence of positrons and electron spin. In his Nobel Lecture, Dirac speculated on the existence of a completely new universe made out of anti-matter! Carl Anderson was awarded the 1936 Nobel Prize in physics for his discovery of the positron.

2.2.7 Mirror Symmetry in Molecular Biology

A variety of symmetries in biology exist. We show three examples of mirror symmetry in Fig. 2.2. Stereoisomers are compounds that have the same empirical and structural formulas but differ in the three-dimensional arrangement of the atoms. In one important form of stereoisomerism two molecules have the same structural formula but one is the mirror image of the other (enantiomers). The existence in nature of left- and right-handed molecules was suggested by observations of the rotation of polarization planes. Amino acid molecules in human cells are always left handed. The right-rotating enantiomer is labeled D-, and the left-rotating, L-. Our two hands are in mirror symmetry and both exist. The wings of the butterfly moth not only show mirror symmetry but also flap symmetrically. These are physical symmetries.

Symmetrical proteins play an important role in molecular biology. The evolutionary selection of certain symmetrical protein complexes appears to be driven by functional, genetic, and physicochemical needs. For example, symmetry affects the flexibility and rigidity of proteins and thus their function. Some symmetric proteins have a preference to have either no ligand bound at all or an entire 'symmetric orbit'

2.3 Arithmetization of Cell Theory

L-alanine D-alanine   Our hands are "enantiomers".   Butterfly moth SATURNIA PAVONIA
                       Sketch by M. C. Escher.

Fig. 2.2 Examples of mirror symmetry in biology. *Source* (Middle) Drawing Hands. https://en.wikipedia.org/wiki/File:DrawingHands.jpg (Fair use of low resolution reproduction). (Right) Jean-Pierre Hamon (1991), https://commons.wikimedia.org/wiki/File:20_petit_paon_de_nuit.jpg

of ligands bound. That is, in some symmetric proteins, the binding of one ligand at some site of the protein quickly leads to the binding of further ligands at the corresponding symmetric sites, so that the resulting structure exhibits the same symmetry as the original protein. Some symmetries have a significant impact on the flexibility of molecular structures.[27] A better understanding of symmetries can therefore lead to better methods of drug design.

2.3 Arithmetization of Cell Theory

In pursuing any intelligent activity, one must always bear in mind Gödel's theorems[28]:

1. Any consistent formal theory capable of producing arithmetical truths cannot be both consistent and complete.
2. Any formal theory containing the truths of arithmetic and truths about provability has a statement of its own consistency if and only if the theory is inconsistent.

In any intelligent activity, we are plagued by the fact that truth is not knowable. In trying to understand things rationally, we constantly fight doubts about the consistency and completeness of the axiomatic system we rely upon. We see certain meanings as obvious in a string of rule-governed symbols without even realizing that we have instinctively found an isomorphism (or perhaps we are genetically coded to find) between the string and things in the real world. This happens most frequently when we communicate in a natural language where we instinctively attribute meaning to words in themselves (and even multiple meanings in a pun). The discovery of new implicit meanings is far tougher, and this requires intelligence. Such discoveries create "aha!" moments in life. Implicit meanings are difficult to discover because they lie hidden in layers of meaning, i.e., it involves a chain of isomorphisms. This

---

[27] See, e.g., Schulze et al. (2014).
[28] Gödel (1931).

is the case in finding meaning that lies hidden in cellular DNA. When AI begins to function at the level of discovering implicit meanings, it would have come of age.

A simple two-layer isomorphism shows why intelligence is required. Consider the digital transmission of music. It has two layers: (1) the isomorphism between arbitrary binary string patterns and sound waves in the air and (2) the isomorphism between sound waves and the electromagnetic waves encoding the music emanated by the transmitter. For one who does not know the technology (e.g., Mr. Average), he will almost certainly never figure out that an isomorphism exists between a binary string and the music he hears. It will also perhaps never occur to him that any binary string can be mapped to an infinite number of isomorphic meanings (although this is obvious to anyone skilled in digital technology) but finding the meaning that fits a context is where intelligence is required, i.e., it requires hypothesis formation using intelligence. First he must determine if the string in encrypted before he proceeds further!

These difficulties apart, an added complication arises when the levels form a loop, i.e., the last level provides input to the first level (it becomes much worse when intermediate levels also loop to provide inputs to previous levels). In engineering, these are called feedback loops. Such loops are used to auto-regulate the behavior of systems in which they appear. However, such loops also constrain the operating envelope of the system, outside of which the system's behavior usually becomes unpredictable, uncontrollable, and even unsafe.

As the number of loops increase, finding a suitable axiomatic system becomes sharply more difficult. Mathematicians have a pet analogy. Given a set of prime numbers you simply multiply them (a trivial, sequential task) to find the corresponding composite number. But given a composite number, searching for its corresponding set of prime factors (even though it is a sequential task) is like finding a needle in a haystack. That is, bigger the composite number, the more horrendous is the search for its prime factors. For example, if the composite number comprises a thousand digits, you might as well give up if all you have is a Turing machine at your disposal. Likewise, discovering the axioms (prime factors) of a non-trivial axiom system (composite number) requires true intelligence. Present-day molecular biologists are desperately in search of an axiomatic system. So far what they have is only a glimpse of the tip of an iceberg, as in Table 2.1.[29] Once we discover successive layers of the isomorphism described in Table 2.1, molecular biology would have made a spectacular advance and we will be designing living organisms according to our fancy. Whether that will be good or bad, we do not know and will never know.

---

[29] Hofstadter (1979), pp. 85, 536.

Table 2.1 Arithmetization of cell theory

Cell	Axiomatic system for number theory
Perfect cell	Complete system for number theory
Strand of DNA	String of the formal system
Strand of DNA reproducible by a given cell	Theorem of the axiomatic system
Strand of DNA unreproducible by that cell	Non-theorem of the axiomatic system
Process of transcription of DNA onto mRNA	Process of interpreting theorem of the system
Strand of mRNA	Interpreted theorem of the system
Translation of mRNA to protein	Mapping to true statement of number theory
Genetic code	Gödel's numbering system
Destruction of the cell	Failure of the axiomatic system to decide
Cell for which there exists at least one DNA strand which it cannot reproduce	Strong axiomatic system for number theory
Immunity Theorem. No immune system can protect the body from every true attack. Death transcends life	Gödel's Theorem. No axiomatic system can reproduce every true statement as a theorem; it must remain incomplete. Truth transcends theoremhood

Adapted from Hofstadter (1979), p. 85, 536

2.4 AI and Its Environment

When developing AI we have a serious problem. We experience the universe, but we have no clue as to how big it is, where it came from, where it is going, or even why it exists. Some religions claim to know the answer (they obviously don't), but physicists have no clue and are ethically frank about it. We have a deep-rooted, compulsive desire to know the universe but all we have are some conjectures that appear to be true and some that are deeply baffling. When we talk of intelligence we must bear this in mind.

2.4.1 The Brain

Inspired by the physicists' standard model of particle physics, biologists too have embarked on a standard model of the brain. On 19 September 2017, the International Brain Lab (IBL) was launched that placed 21 neuroscience laboratories in the USA and Europe into a giant collaborative project. It will focus on how the brain works by focusing on a behavior shared by all animals: foraging.[30] Currently there are

---

[30] Abbott (2017). See IBL (2017). The effort will be to "develop a standardized mouse decision-making behavior, to make coordinated measurements of neural activity across the mouse brain, and to use theory and analyses to uncover the neural computations that support decision-making.".

two large-scale neuroscience projects in place: (1) in 2013, the European Commission announced the 10-year Human Brain Project,[31] estimated to cost more than €1 billion ($1.1 billion); and (2) in 2014, the US Brain Initiative (Brain Research through Advancing Innovative Neurotechnologies (BRAIN) Initiative)[32] was launched to develop neurotechnologies, with $110 million of funding that year. The project is continuing. Other notable efforts include the Blue Brain Project (founded by Professor Henry Markram in May 2005 at the École Polytechnique Fédérale de Lausanne)[33] and the China Brain Project[34] (2018). Google Brain (2011) is a non-biological initiative by Google in AI.[35] The intent to form an International Brain Initiative (IBI) was declared on 08 December 2017.[36] "Since then, the IBI has established a shared vision and aspirational goals, a governance structure, topical working groups, and a 5-year strategic plan. The initiatives and organizations involved in the IBI aim to provide a robust forum for global information sharing and resources for collaborations."[37]

Only about 2% of the body's mass comes from the brain, yet it consumes about 20% of our total energy. In newborns, the brain consumes about 65% of the baby's energy. About 80% of our genes are coded for the brain. An estimated 100 billion neurons reside in our brain. This neural network contains exponential amount of neural connections and pathways.[38] We now know that specific areas of the brain are mapped to certain behaviors. For example, there is a clear link between behavioral problems, e.g., speech and language impairment, and damage to specific regions of the brain. We also know that those with damage to the left temporal lobe can understand speech but cannot say anything or else they drop many words when speaking (Broca's aphasia). On the other hand, damage to a slightly different area of the left temporal lobe leads to the opposite problem, i.e., such people can articulate

---

See also: IBI (2020). "The intent to form an International Brain Initiative (IBI) was declared [on 08] December 2017. Since then, the IBI has established a shared vision and aspirational goals, a governance structure, topical working groups, and a 5-year strategic plan. The initiatives and organizations involved in the IBI aim to provide a robust forum for global information sharing and resources for collaborations." For IBI activities update, visit https://www.internationalbraininitiative.org/about-us

[31] HBP (n.d.).

[32] NIH (n.d.).

[33] Markram (2006). See also: EurekAlert (2018). "Like 'going from hand-drawn maps to Google Earth,' the Blue Brain Cell Atlas allows anyone to visualize every region in the mouse brain, cell-by-cell—and freely download data for new analyses and modelling." Link to the original research article: https://www.frontiersin.org/articles/10.3389/fninf.2018.00084/full. See also: Blue Brain Cell Atlas. https://portal.bluebrain.epfl.ch/resources/models/cell-atlas/ Provides the first digital 3D cell atlas of the whole mouse brain. "Users can view and download the number, major types and 3D positions of all neurons and glia in all 737 areas of the mouse brain.".

[34] Cyranoski (2018).

[35] Our mission. Google AI, https://web.archive.org/web/20180620051038/https://ai.google/research/teams/brain/our-mission/. See also: Dean and Ng (2012).

[36] IBI (n.d.). (Accessed 25 August 2021).

[37] IBI (2020). For updates on IBI, visit https://www.internationalbraininitiative.org/about-us.

[38] Kaku (2014), p. 4.

clearly, but they cannot understand written or spoken speech. We also know that the brain is basically electrical in nature and that there are electrical pathways connecting the brain to the body. It is through electrical stimulation of the brain that we know that the left hemisphere of the brain controls the right side of the body, and vice versa. Since the brain has no pain sensors, observing the inner workings of the brain is comparatively easier to study invasively. This has enabled a fairly accurate one-to-one mapping between specific regions of the cortex and the human body in terms of function and their respective importance. For example, considerable brain power is devoted to controlling our hands and mouth because they are so vital to our survival. We also know that by stimulating parts of the temporal lobe, people can recall, with crystal clarity, long forgotten memories.[39]

The outer layer of our brain, the neocortex, is divided into four highly developed lobes, each devoted to processing signals from our senses, except for the frontal lobe, located behind the forehead (see Fig. 2.3). The prefrontal cortex, i.e., the foremost part of the frontal lobe processes most of our rational thoughts. Damage to this area can impair our ability to plan or contemplate the future. The parietal lobe sits at the top of our brains. The right hemisphere controls sensory attention and body image; the left controls skilled movements and certain aspects of language. If damaged, it can cause difficulty in locating parts of our own body. The occipital lobe is located at the very back of the brain. It processes visual information from the eyes; hence any damage to it can cause blindness and visual impairment. The temporal lobe controls language (on the left side only), as well as the visual recognition of faces and certain emotional feelings. Any damage can leave us speechless or with the loss of recognition of familiar faces.

To understand intelligence, we need to understand how the brain uses chemistry to do arithmetic. In the past, if the brain was rote educated to do arithmetic it was deemed to be doing intelligent work. When Alan Turing in 1936 showed, by mathematical reasoning, that arithmetic can be mechanized,[40] doing calculations became an unintelligent activity. However, conceiving a computing machine required intelligence. Man had earlier learnt to build airplanes not by imitating birds but by understanding the physics of fluid flows. Here too man built computers not by imitating the brain but by understanding the flow of electrons. Electronic computers are exceptionally superior to any human in calculation both in terms of speed and accuracy. In either example we still have scope for further spectacular progress. If we are to create AI far surpassing human intelligence, instead of mimicking the biochemistry of the brain, we need to look at mathematical modeling for inspiration and ignore such activities of the brain that either hinder intelligent activity or do not contribute to it.

In one evolutionary perspective, the human brain's organizing structure broadly evolved to now comprise the reptilian brain, the mammalian brain, and the human brain (see Fig. 2.4).

---

[39] Kaku (2014), pp. 14–16.
[40] Turing (1936).

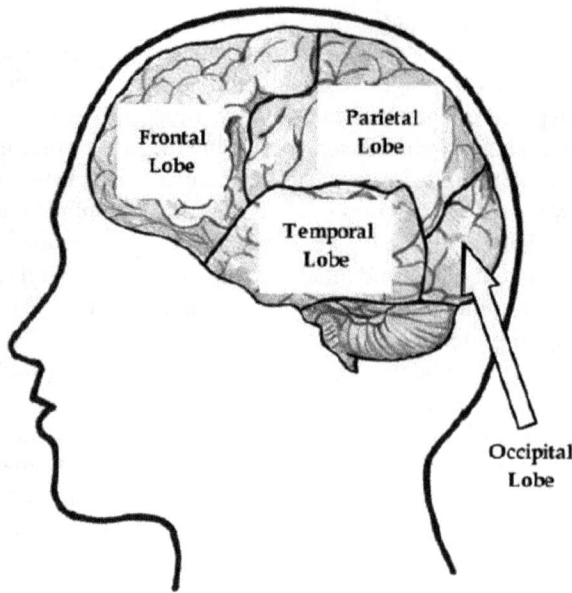

Fig. 2.3 Four lobes of the neocortex. *Source* Alfred (2006), p. 3

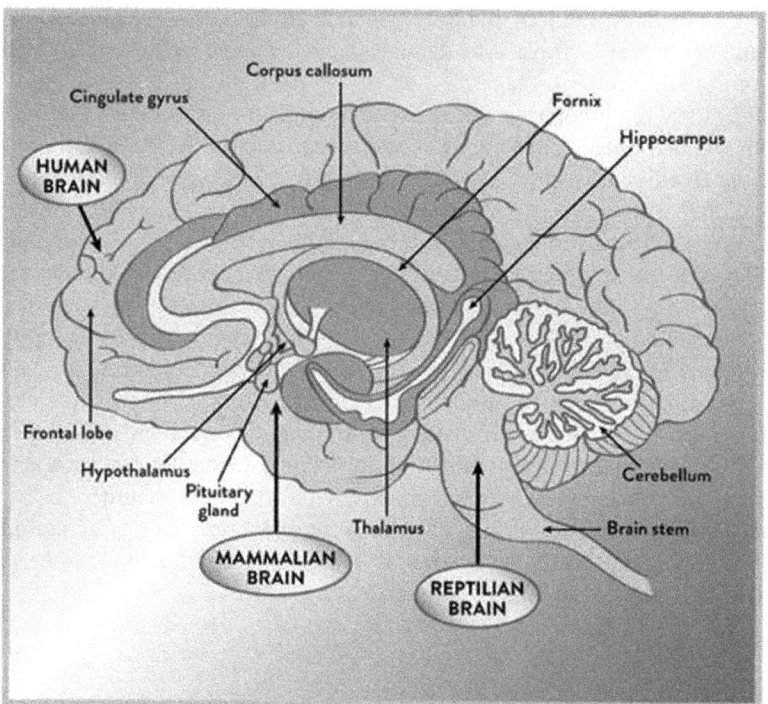

Fig. 2.4 Evolutionary history of the brain. *Source of figure* Kaku (2014). Chapter 1

2.4 AI and Its Environment

(1) Reptilian brain—the back and central part of our brain comprising the brain stem, cerebellum, and basal ganglia. Its evolution goes back to about 500 million years. It governs basic animal functions such as, balance, breathing, digestion, heartbeat, and blood pressure; it also controls behavior, e.g., fighting, hunting, mating, and territoriality necessary for survival and reproduction.

(2) Mammalian brain or the limbic system—located near the center of the brain and surrounding parts of the reptilian brain. The limbic system deals with emotions. It is prominent among animals living in social groups, e.g., the apes. It deals with problems of social dynamics, e.g., sorting out potential enemies, allies and rivals. Parts of the limbic system that control behavior are the hippocampus, amygdala, thalamus, and hypothalamus (named to indicate that it lies below the thalamus). Their respective functions are:

- Hippocampus. It is the gateway to memory where short-term memories are processed into long-term memories. If injured, the ability of the hippocampus to create long-term memories is lost. You then live only in the present.
- Amygdala. It is here that emotions, such as fear are first registered and generated.
- Thalamus. The name means "inner chamber". It is a large egg-shaped mass of gray matter, having a small amount of white matter. Like a relay station it gathers sensory signals from the brain stem and sends out to various cortices (the outer regions of an organ or structure). It plays an important role in controlling the emotional tone of a person.
- Hypothalamus. It regulates the body temperature, our circadian rhythm, hunger and thirst, and certain aspects of reproduction and pleasure.

(3) Human brain. It is the most recent region of the mammalian brain—the cerebral cortex. It is the outer layer of the brain. The most recent evolutionary structure inside the cerebral cortex is the neocortex. It governs higher cognitive behavior. It is most developed in humans. It is as thick as a napkin yet is comprises 80% of the brain's mass. In humans the neocortex is a highly convoluted surface which means a very large amount of surface area is crammed into the human skull.

Since the 1990s, new experimental techniques have provided us with much more detailed knowledge of the brain. Within a span of 15 years, a wide range of scanning techniques were invented using the electromagnetic force. It provides the great advantage that electromagnetic radiation can pass right through tissue without doing damage. In fact we are now able to reveal the mechanics of thought! The techniques used are magnetic resonance imaging (MRI), electroencephalogram (EEG) scans, positron emission topography (PET; this uses the weak nuclear force and not the electromagnetic force) scans, magnetism in the brain, deep brain stimulation, optogenetics, transparent brain, four fundamental forces, etc. Spatial and temporal resolution of images is improving continuously, and the cost of imaging is reducing as is the size of the equipment. It is now possible to look at minute parts of the brain and make precise measurements and see the inner workings of the brain in 3-dimensions.

Population of the Earth (in billion markers)			
Year	Population	Year	Population
1804	1 billion	1999	6 billion
1927	2 billion	2012	7 billion
1960	3 billion	2024	8 billion
1974	4 billion	2048	9 billion
1987	5 billion		

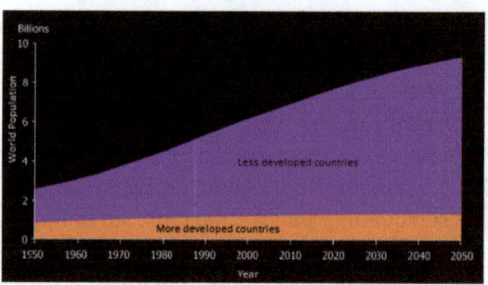

Fig. 2.5 Population of the Earth. (Left) World population growth. (Right) World Population Growth, 1950–2050. *Source World Population Prospects: The 2010 Revision*, United Nations, 2011, http://www.un.org/en/development/desa/population/publications/pdf/trends/WPP2010/WPP 2010_Volume-I_Comprehensive-Tables.pdf. Note the rapidly increasing population size in the less developed countries

Eventually the human brain evolved to believe in gods. Gods, emotion, and illogic have no place in AI. This by itself should serve as a warning that blind mimicking of the biological brain may lead to failed AI. For example, much of the euphoria with deep learning is now fading because a fundamental understanding of deep learning is still missing.[41]

2.4.2 Growing Human Population

When the Industrial Revolution began around 1760 the world's population was less than a billion. It reached a billion in 1804. By 1960 it had crossed 3 billion, and by 2000 it had crossed 6 billion (see Fig. 2.5). In 2023 it crossed 8 billion. In our present world, "from each according to his ability, to each according to his need" is no longer a sustainable dogma. Those in need are rapidly rising in numbers, while the capacity to practice altruism by the rest is not. Even in a world of abundance, those in need will starve if they lack purchasing power. In the AI-dominated world, it is "to each according to his plus his benefactor's ability". This is obvious from present population growth trends.

To understand the gravity of rising population, one must get a historical perspective. We the *Homo sapiens* have been around for about 300,000 years.[42] Till about 12,000 years ago we remained nomadic hunter-gatherers before settling down to a pastoral-agricultural life of rearing animals and sowing seeds on Earth. Society structured itself into families; women took care of the household, and men earned a livelihood.

---

[41] See, e.g., Woodie (2017). See also: Somers (2017) and Elsayed et al. (2018).

[42] SNMNH (2018). *See also* the original papers announcing the discovery in Nature: Hublin et al. (2017), and Richter et al. (2017). Prior to these papers, *Homo sapiens* were said to have been around for about 200,000 years.

Around 1500 AD, another socio-economic transition took place that lasted till the later half of twentieth century. During this period an industrial economy developed with increasing growth of industrial activity and mechanization of agriculture. Within five centuries, the economy advanced from using animal power to steam power, to fossil fuel power, to electrical power. Along with each new source of power, society too restructured itself into increasingly complex communities—extended families, cities, nations, alliances, institutions, modes of governance, dominions, etc.—and economies that ranged from family businesses run locally to multinational corporations operating globally and employing millions of men and women. The woman's role in society too expanded into areas that were earlier exclusive to men whether it be in business, politics, arts, science, etc. Women are no longer tied to home and hearth but an active player competing with men in the wider world of money and power where gender differences are continuously becoming blurred.

Industrialization propelled a sharp rise in productivity via mechanization, employment opportunities, wages, urbanization, urban population density, and above all, another change in lifestyle that expanded the middle class rather rapidly. Compared to agriculture, manufacturing requires far less land; its growth limiting factors are skilled labor and capital. Skilled labor is much more scalable through training than finding cultivable land. Increasing population density led to more efficient capital markets (partly due to closer proximity of transacting parties), which further made financing of manufacturing capacity easier leading to further increases in output. Around mid-nineteenth century, in Europe and the USA, this upward spiral of economic growth got a further boost from advances in technology that led to more wealth, more capital formation, still more technological progress, in an *apparently* self-sustaining, gravity defying, upward flight.[43] It saw airplanes, rockets, and spacecraft!

Until the fifteenth century, progress in Europe had substantially depended on transfer of technology from Asia and the Arab world. But in the sixteenth and seventeenth centuries, science in Europe made amazing progress due to a galaxy of scientists that included Copernicus, Erasmus, Bacon, Galileo, Hobbes, Descartes, Petty, Leibnitz, Huygens, Halley, and Newton. From Newton onward advances in science and technology took a meteoric path.[44] Rational scientific thought gained deep roots and eventually destroyed the tyranny of the Catholic Church. Along with science grew great universities, research laboratories, and manufacturing industries dependent on technologies derived from scientific knowledge. It is now evident that highly industrialized nations became so not because of science alone but because they also assiduously built the "vital underlying institutions of property rights, scientific enquiry, and capital markets."[45] Over time, these factors have become encoded in their culture and broke their chains of poverty. The industrial stage emphatically showed that recovery from disaster, such as World Wars, is faster and surer when property rights are guaranteed.

Of the industrial era, in 1981, historian Daniel R. Headrick wrote:

---

[43] Bernstein (2004).
[44] OECD (n.d.).
[45] Bernstein (2004).

Western industrial technology has transformed the world more than any leader, religion, revolution, or war. Nowadays only a handful of people in the most remote corners of the earth survive with their lives unaltered by industrial products. The conquest of the non-Western world by Western industrial technology still proceeds unabated.[46]

At the time, nothing perhaps underscored the remarkable breadth of human ingenuity in developing technology than accomplishing the safe landing and walking on the moon by two US astronauts, Neil Armstrong and Buzz Aldrin, on 20 July 1969, and their subsequent safe return to Earth on 24 July 1969 with the event being broadcast live on TV to a global audience. In Adam Smith's days (1723–1790) such a feat was possible only in science fiction, fairy tales, and mythology. Since late 1980, technology has advanced and expanded at a remarkable pace, scope, novelty, variety, and scale. Advances in automation, electronics, communications, biotechnology, superconductivity, computing, plastics, and more have fundamentally changed society including the goals, ambitions, and lifestyles of the masses. Now trillions of dollars, millions of jobs, and geopolitical power flow from the exploitation of science-rooted technologies rather than from raw materials and smoke belching factories. In 1990, Erich Bloch, a former head of the National Science Foundation in the USA aptly summarized the role of technology in the marketplace at the time:

> In the modern market place, knowledge is the critical asset. It is as important a commodity as the access to natural resources or to a low-skilled labor market was in the past. Knowledge has given birth to vast new industries, particularly those based on computers, semiconductors, biotechnology and designed materials.[47]

The economy dealt mainly with physical stuff—chemicals, steel, minerals, etc. The knowledge worker, though important, was more keen to shape the future rather than the immediate present. The rapid and sustained rise in the world's population since the Industrial Revolution from less than a billion to more than 8 billion as we write (July 2023) was remarkable. It was possible because of rapid increase in scientific knowledge, its application to development of technology, and above all society's ability to educate and skill its people in a continuous stream of sufficient capacity to provide a science and technology workforce required to run a multitude of new and emerging industries. During the industrial era, a self-sustained, increasingly prosperous synergy between man and natural resources developed till about 2000. Now, on the negative side, society is utilizing natural resources at a rate faster than their replenishment rate[48]; the environment is degrading. The hope is that humanity, because of a larger population of S&T trained people, will find solutions before things get out of hand. That appears to be wishful thinking.

The Industrial Revolution showed that science is not just about esoteric theories but also a social enterprise. For those inquisitive about the working of the universe, science is an addiction. The post-industrial society is now primed to be driven by AI. The economy is already being driven by knowledge-intensive services that require

---

[46] Headrick (1981), p. 4.
[47] Bloch (1990), p. 9. Quote as reproduced at Warshofsky (1994).
[48] Alcoforado (2015). See also: WRF (2017).

2.4 AI and Its Environment

even less land and labor than manufacturing. It requires huge capital investment and its appetite for science-rooted innovative technologies is ravenous; its market is truly global. Its Achilles' heel is its peripatetic core STEM talent. The average-skilled workforce in the service sector while initially scalable due to its natural ability to self-learn, learn by association with others in the community, or be trained in large groups, is no longer so. The skill levels required of even entry-level jobs have escalated rapidly, and commensurately available middle-class jobs have diminished. The hitherto rapidly expanding middle class is now equally rapidly downsizing, giving way to machines with superior skills and easy availability in an increasing range of skills. Mr. Average of the middle class is on the verge of losing his identity and his dignity, not just his income. The elite PhD-level workforce too is now in danger. The job market is churning, shrinking, and vaporizing. While machines can be easily fitted with AI, humans cannot (at least not yet). Ironically machines are neither looking for jobs nor in need of one, much less competing for one. They are emotionless, oblivious of the past and the future, without any need for spiritual balms or companionship.

This unintended turn of events has occurred because exceptional men with unbridled curiosity configured machines to compute, while Nature has configured most other men for mere procreation. The previous stages—agricultural and industrial—though each made a dramatic break from the past, carried the prospect of improving humanity's collective lot through adaptation by reskilling. The present break has caused fear of sinking into poverty for want of a job because the scope for reskilling has almost vanished except for the highly intelligent. Skilling people to meet market demand for highly skilled and gifted people with deep interdisciplinary knowledge is no longer scalable because such people must remain on a life-long learning curve to stay competitive. To bridge shortages, they must be imported or poached, usually from developing countries, whose development agenda then takes a hit. Finally, the post-industrial society, to maintain its economic growth, needs an informed citizenry, tight immigration policies, and undoubtedly, new forms of government and societal structure. Inaction, mistiming, and lapses on such matters can easily lead to catastrophes and mayhem if unemployment soars. The palliative for unemployment cannot be unemployment insurance, social security, etc.; it will require novel and perhaps untested means. Immigrants deemed a drain on a country's economy would be shunned. Euthanasia on demand may become morally acceptable in lieu of suicide.

Economic growth in the industrial stage depended on society's ability to create and efficiently use knowledge to produce marketable products and services and turn them into necessities. Physical stuff is subject to the laws of scarcity; prices of material goods depend on demand and supply. Knowledge is intangible, shareable, extendable, creatable, storable, and marketable. It spawns ideas. Modern-day communication services permit instant and global spreading of ideas. This has turbo-charged the global economy by accelerating innovation and their commercialization on a global scale. In the process, the economy has become exceedingly complex and hence fragile.

Economists, as always, were caught unaware by this tectonic shift and the gathering acceleration with which the post-industrial economy began moving. Most of them are still focused on the scarcity and mismatch of physical and human resources while the economy is already under the iron grip of unpredictable, disruptive, breakthrough creativity that comes from brilliant minds. Their earlier theories based on land, labor, capital productivity, and above all their irrational theory of rational expectations and efficient market hypothesis were anyway farcical. Economic growth has always been driven by human imagination and innovators for which no economic theory exists.

In 2008, Ellis Rubenstein wrote:

> [I]n a post-industrial age the keys to economic sustainability for urban centers will be education, science, technology, finance, and a system that stimulates entrepreneurship. Urban centers whose researchers and university administrations remain "siloed"—disconnected from one another, from industry, and from venture capital—will fall behind. The achievement of excellence solely through the global collaborations of individual investigators will no longer guarantee the excellence of an institution, much less the region in which it resides.[49]

All these have come to pass. Upcoming technologies will be even more breathtaking given the R&D strides already made in biotechnology (stem cell manipulations, synthetic DNA, rapid DNA sequencing, etc.), nanotechnology (carbon nanotubes, graphene, etc.), superfast switching of quantum light sources, cloud computing, data analytics, efficient conversion of solar energy to electricity, etc. The source of economic growth is no longer the brawn but the brain, and above all artificial intelligence!

"Necessity is the mother of invention" no longer dominates; invention-driven necessity does. It began with the mobile telephone, the credit card, the Internet, and the Windows operating system—each became a daily necessity. AI robots and humanoids will soon join them to serve those humans lucky enough to have a job.

All this while population density on Earth has been increasing and stressing and straining the Earth's natural resources by consuming them faster than their natural replenishment rate. Human intelligence was used to increase consumption while ignoring the replenishment problem. It has now come to haunt us. If only the economists had studied the logistic map in chaos theory[50] and the Erdös-Renyi theorem in graph theory[51] they might have seen that the secret of socio-economic

---

[49] Rubenstein (2008).

[50] May (1976). In this paper, May made the insightful remark, "Not only in research, but also in everyday world of politics and economics, we would all be better off if more people realized that simple nonlinear systems do not necessarily possess simple dynamical properties." Robert McCredie May (1936–2020) was President of the Royal Society (2000–2005). He died on 28 April 2020.

[51] Erdős and Rényi (1960). It is an amazing result in mathematics. It is the basic means by which men, machines, ideas, atoms, or anything for that matter that may randomly connect with each other, aggregate into societies, molecules, clubs, etc. The paper has been cited in mathematics more than 17,000 times.

2.4 AI and Its Environment

dynamics lies hidden in them.[52] They provide insights about large-scale socio-economic dynamics—an invaluable aid to policy planning for governments. The Erdös-Renyi theorem tells us that when socio-economic networks undergo a change, society splits and reorganizes into new communities with shared interests, advantages, and woes. And this happens rather spontaneously and pseudo-randomly and the distribution of wealth between the rich and the poor fluctuates roughly according to the logistic map. The faster technology advances, the more susceptible will the socio-economic fortunes of the world become to rapid fluctuations and possibilities of chaos. This knowledge by itself forewarns and therefore forearms policymakers about approaching risks and provides some extra time to evaluate available means of dealing with them. Even when chaos abounds and all appears lost, the logistic map tells us there may still be oases of calm (before an impending storm) to be exploited by the foresighted.

Chaos in the logistic map was a major discovery in 20th-century theoretical physics. The other outstanding discovery that also involved random outcomes was quantum mechanics. Both chaos theory and quantum theory are at the cutting edge of research with tremendous potential for creating new technologies. There is also an intriguing relationship between the two theories, and the distribution of prime numbers in number theory, generally referred to as quantum chaos,[53] which has a bearing on how future computer chips may be designed. This immediately tells us the nature of technology advances AI will introduce in the near future.[54] This means that rote education is on its way to becoming history.[55]

Lessons learnt along the way

In terms of population growth rate, the world has seen three distinct phases. The first, premodernity, was characterized by very slow growth rate with equally low economic growth rate; the second, beginning with the onset of modernity, was characterized by, relative to the first, rapid increase in population yet supported by rising standards of living and improved health and longevity due to some amazing advances in science and technology. We are now in the third phase with an increasing population, a muted population growth rate and a downward trend in the creation of stable job opportunities while S&T advances continue to follow an exponential trajectory powered by AI and solar energy.

Generally, civil society establishes itself around a regulated set of social structures. Each structure stabilizes itself into a distinctive arrangement of institutions to facilitate and maintain human activities in myriad ways. *Inter alia*, through tradition, custom, and law, societies set up institutions for sexual reproduction, the care and education of the young, opportunities for gainful employment, and the care of the elderly in which marriage and kinship play important roles. For most of history,

---

[52] Bera (2021a).
[53] Cipra (1999).
[54] Bera (2019a).
[55] Bera (2021b).

society gradually developed technology to avert risk to life and limb and the instinctive need to survive and propagate the species. Technology development began to rise sharply coinciding with the birth of the millennials.

A distinctive aspect of emerging technologies is their ability to create necessities or a 'must possess feel' not felt before. This has led to aspiration-driven marketing. Much of this is visible in myriad digital communication-plus devices ubiquitously available and affordable. Now instant communication links that connect humans and devices via the Internet of Things (IoT) are increasingly taken for granted. It has set in motion a disruptive restructuring of society by an "unseen hand" into a malleable global structure where people are tagged with a profile matrix that includes lifestyle, nationality, education, employability, religion, etc., usually in that order of importance. Society increasingly celebrates the individual than the family. Weakened family ties amplify the mental stress of unemployed individuals and prod them to reexamine their religious beliefs usually inherited from the family. The first millennials were born in the incipient stages of this disruption. The rapid changes impacted family structure and lifestyle, employment, skilling and reskilling opportunities caused by the automation of many hitherto human activities (physical and mental). In parallel, the welfare management of a growing population of retirees, culturally alienated from their millennial progenies, who too found themselves facing an uncertain and unpredictable job market, became a new cause of health and welfare worries. The old socio-economic structure began crumbling, and a new stable structure is yet to take shape.

The times when monarchs or governments could change things by diktat are over. Today a tiny group with disruptive ideas can create a start-up that can not only change a country but the entire world. At a young age they become flush with wealth. The world is flush with youngsters who have a deep hunger to learn and forge dynamic partnerships to rapidly expand their enterprises. Only a few will succeed. The world now moves on the power of ideas growing grander by the day. Archaic traditions and conventions are passé. The Industrial Revolution made human brawn power obsolete; the post-Industrial Revolution has made rote education obsolete. AI machines will replace every rote educated human. The key to human survival is to improvise and quickly sunset habits, institutions, and practices of the past that can no longer harmoniously blend with the future. For individuals, the problem is identifying and deciding what to sunset because the future is unknowable, and individual talent will often be over-shadowed by AI. Emerging technologies now shaping humanity's future include AI and robotics, novel modes of transport and renewable power, data mining, and data privacy. Their impact will be huge, presently unpredictable, and unforeseeable.

2.4.3 *Exponentially Accelerating Technologies*

In 2001, Ray Kurzweil identified genetics, nanotechnology, and robotics as the three overlapping revolutions which will define the future.[56] Genetics will let us reprogram and even create novel biological entities (the intersection of information and biology); nanotechnology to manipulate matter at molecular and atomic scales (the intersection of technology and physics); and robotics to create artificial intelligence (AI, the building of strong AI on an information base). The integrator is information-driven technology. Kurzweil expects it to occur within three decades (starting from 2001), that is, during the life time of the millennials. Note that till now humanity had assumed intelligence is its most important and powerful attribute. With rapid advances in AI, that assumption now appears to be deeply flawed.

Kurzweil estimates that biotechnology is advancing at an exponential rate, paralleling those in information technology (IT) and IT-enabled technologies. Inventors are now poised to bring ever more powerful technologies faster than we naively imagine or can anticipate because we are mentally wired to extrapolate linearly rather than exponentially; nonlinear extrapolation is not a part of our intuition. It is extremely difficult to think of change in exponential terms (see Fig. 2.6, left). We are now on the knee of the curve in terms of the rate at which technology is advancing. Phase transition (i.e., a transition between distinctly different physical states (phases) of the same substance or entity; see Fig. 2.6, right) is ubiquitous in Nature, *e.g.*, when steam condenses to water suddenly at 100 °C at atmospheric pressure due to the random linking of water molecules through hydrogen bonding. Phase transition immediately gives us the basic *involuntary* mechanism by which a society spontaneously reorganizes itself into a new socio-economic structure as people, machines, resources, etc. connect or disconnect with each other randomly, say, in the face of a sudden disaster. Phase transition is highly pronounced in an Internet of Things (IoT) connected world where one can see how people are spontaneously polarized on issues that concern them via social media. The Erdös-Renyi theorem says polarization will occur, like it or not.

2.4.4 *Know the Power of Your AI Competitor*

The most remarkable change in technology in the past century has been in computing. Per US dollar, you can get about 10^{18} times or more computations done today than in 1900! (see Fig. 2.7). A defining feature of an advanced society is how it communicates with humans, machines, and institutions, the means by which humans control and coordinate strategy.

Communication channels for the millennials expanded suddenly with the invention of the microchip in 1959 (a product of the electronics revolution) that enabled the

---

[56] Kurzweil (2001). See also: McShane and Dorrier (2016).

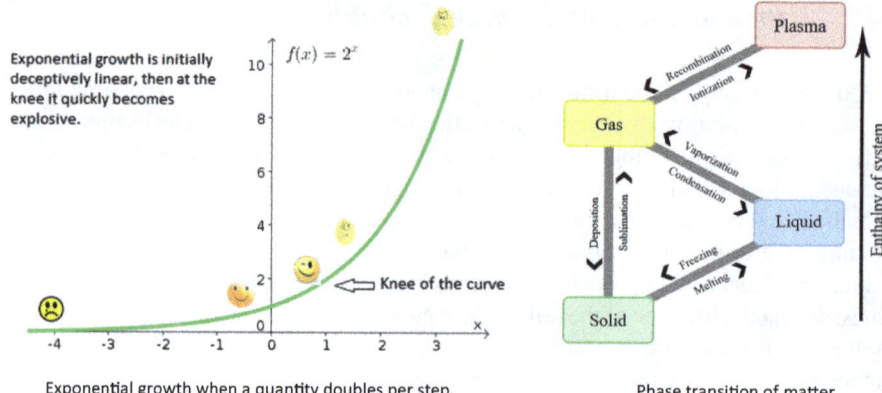

Fig. 2.6 Exponential growth and phase transformation. *Source of figures* (Left) Author. (Right). https://upload.wikimedia.org/wikipedia/commons/0/0b/Phase_change_-_en.svg (in Public domain)

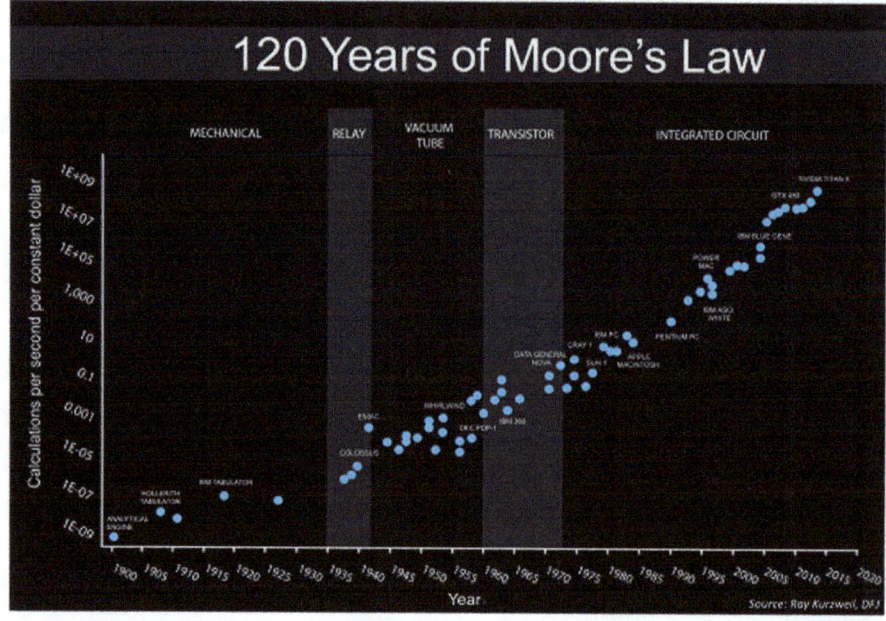

Fig. 2.7 120 years of Moore's law. *Source of figure*: Steve Jurvetson. File: Moore's Law over 120 Years.png. An updated version of Moore's Law over 120 Years (based on Kurzweil's graph). Wikimedia Commons. https://commons.wikimedia.org/wiki/File:Moore%27s_Law_over_120_Years.png

2.4 AI and Its Environment

personal computer (in 1975 as kits) and eventually the present day ubiquitous smartphone that include high-speed mobile broadband 5G, motion sensors, camera, and mobile payment features that fit into a shirt pocket. Consequently, the world of the millennials and their successors is much less human-centric than machine-centric. STEM educated humans need not just a language with which to express myths but also logos, *i.e.*, a language with which to reason in accountable writing and speech. This is fulfilled, to a good degree, in esoteric (and hence limited to relevant experts) modern scientific languages, by heavily depending on the language of mathematics. Scientific languages speak with "the highest, universally binding authority, worldwide about everything in the world" and its "authority is fundamentally egalitarian and democratic; for it and with respect to it, nothing counts but 'the non-violent force of the better argument' (Jürgen Habermas)."[57] This, of course, does not eliminate the criticism that scientific languages also narrow man's perception of reality to what can be expressed in scientific language. Our successor species may evolve to deal with this problem better because their survival may depend on it.

The electronics revolution was ignited by developments in human–computer interface, networking of digital machines, and microchips. In the millennials' world, man–machine interaction and action-at-a-distance with the speed of light is taken for granted. That is only the beginning. Globally connected digital devices and the Internet of Things[58] (IoT), with each device and service becoming smarter by the day with embedded AI that include cognitive functions, are on their way to becoming ubiquitous. Along with it, human–computer interactions are becoming more like human–human interactions because AI has now made significant advances in mimicking cognitive functions. This raises the specter of *en masse* retirement of humans even from "intelligent" tasks. Technology and society are now so intricately interwoven that without understanding their relationship in terms of information flows, humans may inadvertently place themselves in extreme danger of becoming unemployable. For developing countries like India, the danger is even more.[59] Emerging technologies inspired by the likes of IBM Watson,[60] Google's AlphaGo,[61] and Carnegie Mellon University's Libratus[62] clearly indicate such a possibility is round the corner. On another important front, AI may provide an amazing service that of peer-review of scientific papers, in the near future.[63] When such a feat is accomplished, it will be but another step for robots writing papers thus making even human researchers redundant, inefficient, and obsolete. As Janne Hukkinen noted, "New knowledge which humans no longer experience as something they themselves

---

[57] Weiß and Schwietring (2018).
[58] IoT is the network of physical objects—devices, vehicles, buildings, etc., which are embedded with electronics, software, sensors, and network connectivity, which enable these objects to collect and exchange data.
[59] Yahoo (2016).
[60] Best (2013), Young (2016).
[61] Gibney (2016).
[62] Solon (2017).
[63] *See, e.g.*, GenomeWeb (2017). *See also*: Stockton (2017).

have produced would shake the foundations of human culture."[64] Greater the sophistication in communication, the more elaborate will be its language, especially its associated grammar. The post-industrial economy will depend on machine interpretable, error-free communications.

AI and intellectual property rights

The impact of AI on intellectual property rights (IPR) regimes is now a serious concern.[65] Historically, these regimes arose to facilitate, indeed drive, economic growth as a means of improving people's well-being by equitably rewarding inventors. By law, patents can be granted to only human inventors of a novel, useful, and non-obvious invention, and they must describe their invention in writing (written description) with clarity and in sufficient detail so that others knowledgeable in the arts related to the invention can reproduce it independently. Implementations that can, in principle, be implemented by a mind alone are not patentable. Thus inventions produced by AI machines are not patentable on several counts. Machines are not humans. Further, AI-created inventions will be logically and mechanically derivable theorems from some axiomatic system the AI machine is embedded with. Thus the entire population of similar machines or their clones will be deemed capable of ab initio reproducing the invention on demand if called upon to do so. Thus the given invention will not only belong to the prior art (the axiomatic system) but it will also be obvious because it can be invented without any application of mind (call it machine instinct!). Furthermore, it may often become possible to claim non-infringement because the contested patented invention would have been obvious based on prior art embedded in a contemporaneous AI machine. In short, a vast range of inventions will spring forth from AI machines on demand without violating the laws of Nature or of man-made patent regimes. Thus, such studies as conducted by the National Academies (of the USA) may well be futile efforts in 'Advancing Concepts and Models for Measuring Innovation'.[66]

The IT revolution has unevenly affected the world in terms of enhancing living standards, polarizing wealth distribution, governance, the economy, employability, etc. Successful diffusion and adoption of IT is capital intensive. It has enlarged productivity gaps between frontier firms and others. Income and wealth inequality has increased. In the past two decades, the top 1% have gained enormously, while the bottom 80% have steadily lost. The job-mix in the economy continues to churn faster than peoples' ability to adapt since many routine tasks are being automated and skill levels required for new jobs is steadily rising. There is great uncertainty about future wage-earning jobs—what will they be, how long will they last, what skills will they require, what future prospects will they carry, where will they be available, how can one reskill, etc. In fact, how the world will reorganize itself is not at all clear. There will be phase transitions galore at different skill levels for humans. Gathering and analyzing data to even discern trends in what is happening

---

[64] As quoted in Stockton (2017).

[65] See, e.g., Bera (2019b). See also Chaps. 4 and 5 of this book.

[66] NAP (23640) (2017). See also: Crawford and Calo (2016), White House (2016a, b, c).

has become impossible because the situation is fluctuating so fast. Statistical analysts are clueless. While new ways of integrating and analyzing data from diverse data sources without breaching privacy and confidentiality could reveal useful information about the changing status of the workforce, it would be useful only as history and not for planning. The situation is heading toward chaos. PhD-level STEM education will significantly influence worker income, but matching education with opportunity will be a loaded game of dice and available only to a few.

In 2017, I and Sunish Raj studied disaster management of human resources. We concluded:

> The principal contributors to the impending disaster is an unavoidable socio-economic phase transition that we infer from a fundamental result in graph theory in mathematics and the magnitude of the disaster is estimated based on the logistic map in chaos theory. The scope of managing the disaster will be extremely limited and restricted largely to the survival of the creatively intelligent among the human species. This disaster will be global, its scale will be massive, and human society will find itself grossly underprepared to deal with it because of its psychological reluctance to accept it as a possibility.[67]

Why are we so confident? We view the two remarkable, half-century old, mathematical results as laws of Nature. Their mathematical statements appear simple, and what they capture and reveal is profound.

2.4.5 Science Concepts that Presently Drive AI

The conceptual framework around which this book is written heavily draws from the foundational works of the people listed in Table 2.2. AI researchers who care to revisit and refresh their minds with the concepts enunciated by the luminaries listed in Table 2.2 will find conceptual gems they need to advance AI.

2.5 Message for Those About to Step Out of College

> People are always blaming circumstances for what they are. I don't believe in circumstances. The people who get on in this world are the people who get up and look for the circumstances they want, and, if they can't find them, make them. – Mrs. Warren's Profession by George Bernard Shaw (1983)
>
> The reasonable man adapts himself to the world: the unreasonable one persists in trying to adapt the world to himself. Therefore all progress depends on the unreasonable man. – Man and Superman by Bernard Shaw (1903), p. 238.

The millennials are caught in a whirlwind of phase transitions, the essential features of which are captured in the Erdős-Rényi theorem and the logistic map. Rapidly increasing connectivity among men and machines has imposed upon the

---

[67] Bera and Raj (2017).

Table 2.2 My personal list of intelligent people who introduced concepts that drive intelligence

Since Galileo Galilei (1564–1642), scientific & mathematical knowledge has advanced at an accelerated pace. My personal heroes who I see as founding fathers are Galileo, Newton, Maxwell, and Einstein (Classical physics). They told us about force and its effect on movements of matter and energy in the Universe. The other luminaries are
Nöther (Symmetry & conservation laws). She told us the deep meaning hidden in the action principle
Euler (Graph theory). He told us about fundamental pairwise relationships between objects
Planck, Einstein, Bohr, Born, Schrödinger, Heisenberg, Dirac (Quantum mechanics). They showed us the structure of the Universe at sub-microscopic scales and the limits Nature places on the measurements we can make in it. The structure is the Hilbert space
Gödel, Turing, Chaitin (Formal systems). They showed us the limits of reasoning and computing we can bring into our efforts to understand the Universe
Shannon, Landauer (Information theory). They told us the meaning of information and its relation to physics (particularly thermodynamics) in the language of mathematics
Crick and Watson (Molecular biology). They told us how to read the book of life
Mandelbrot (Chaos and fractals). He took us beyond Euclidean geometry. "Clouds are not spheres, mountains are not cones, coastlines are not circles, and bark is not smooth, nor does lightning travel in a straight line"

global socio-politico-economic structure, a series of phase transitions we now witness. More will occur in areas where massive connectivity is the key. Being caught unawares will soon become the norm, encompassing a wide range of activities and issues. It is inevitable because enabling massive general connectivity is now a part of the global infrastructure maintained practically *gratis* by the Internet and the World Wide Web (WWW). On this infrastructure, ride specialized social media, which connect people globally, anytime, anywhere on any issue and in commerce. If the Internet collapses, so will the world.

A generic feature of all phase transitions is that immediately before transition, existing rules of the game begin to crack, and during transition, they break down. Post-transition, new rules must be framed and enforced to establish order. One may not always succeed in doing so. Since phase transition is a statistical phenomenon, the only viable option for managing it is to manage groups and temporarily suspend individual rights. At the human level, this raises conflicts between individual rights and group rights for which temporary equitable conflict resolution mechanisms must be found, often on a case-by-case basis. A phase transition in societal structure caused by AI-powered, large-scale automation, will expose man's inability to cope and align with the profound socio-economic forces unleashed by automation. There is presently a psychological reluctance to accept it as a real possibility. And therein lies grave danger.

Those playing the stock markets are quite familiar with such fluctuation and so are physicists controlling laser intensities.[68] When a resource is conserved, e.g., energy in the universe (law of conservation of energy—kinetic plus potential or momentum

---

[68] Hnilo (1985).

in physics; or money in circulation in the banking system—credit and debit) which is allowed to oscillate between two or more states, then the faster the system oscillates, the more fragile it is likely to become. It is then no longer possible to control the system for the short term, but one may be able to control it for the long term in some coarser sense. This is because all such systems tend to exhibit some large-scale behavioral pattern which, if understood, may be cleverly exploited. Physicists call such patterns "strange attractors". An example is the weather. We can with considerable confidence predict the onset of seasons but not the daily fluctuations in the weather.

The message. Our propensity to be networked with humans and machines has unintended consequences as does our eagerness to utilize conserved resources without allowing the resulting waste (the laws of thermodynamics ensure there will be waste) to be recycled fast enough for use in the next cycle. Since mathematics is abstract, its results apply wherever the abstraction meaningfully fits; it really does not matter what the mathematical variable gives meaning to—inanimate, animate, or abstract entities. In fact, learning, whether by humans or machines, is all about connectivity and rapid changes in connectivity (phase transition) and the rate at which events change and create diversity (the logistic map). When we suddenly understand or discover something, it is due to a phase transition in our mind.

The new generation coming out of universities will henceforth face wild swings in employment opportunities depending on how interdisciplinary knowledge evolves, and how fast they can upgrade their skills. Survival of the fittest now has a new meaning. The new generation is in the throes of a self-referential system that is thrashing. It is self-referential because AI development requires the human mind to understand itself, looping between knowledge and ignorance, at a faster and faster rate. An isomorphic interpretation of the logistic map tells us that this knowledge-ignorance tussle has already entered the wildly oscillating regions of the logistic map, where society can no longer measure, assess, and upgrade the population fast enough to match demand and supply of required skills. Unemployment is surging like a pandemic, and we have no vaccine to control it, much less eradicate it.

STEM has advanced to a level where it is well beyond the comprehension of almost all *Homo sapiens* except for a few. But the axiomatized part of STEM knowledge, in principle, is eminently programmable and hence can be embedded into AI machines. Democracies where masses ignorant of STEM allow equally STEM ignorant politicians to govern them are in danger. This means practically the whole world. Unless one understands axiomatic systems where iterative actions lead to shortening and lengthening of theorem-strings in the derivation process[69] cannot accept the fact that economic ups and downs are inevitable. Any attempt to control them beyond a point will likely aggravate their chaotic behavior. Attempts to reduce poverty increases population size and their longevity. In a resource constrained world this additional consumption then promotes poverty as per the logistic map. It is the interplay between opposites trying to find a stable balance—the back and forth negotiations—that create instabilities that lead to stampedes we call chaos. Such chaos

---

[69] This phenomenon is elaborated in Hofstadter (1979).

subside if there are internal or external mechanisms which, if triggered, will douse the chaos by draining out the energy driving it. For example, violent crowds subside when they are physically exhausted or disperse when superior external force is used. The external force that has now come into play is climate change and a devastating pandemic.

Managing the future thus reduces to managing triggers (phase transitions and bifurcation points in the logistic map). Some triggers are sensitive, e.g., rapid poverty removal programs that create low-skilled jobs but raise aspirations only to be dashed when machines take over; welfare schemes where the net return on investment to society is negative; and education and skilling programs that fail to fulfill society's demand for creative knowledge workers. When people have a lifestyle, they want respect and freedom and dominance over others. When they become unemployable, they will willingly trade freedom and self-respect for employment, almost any employment. Management of human resources in a transitional period is basically a game of chance. We are experiencing the birth pangs of a new era in the lives of the *Homo sapiens*. We are being driven by AI, robotics, and automation and experiencing the dramatic changes they have enforced on industry, socio-economic fundamentals, and our understanding of what it means to compete. The central lesson we have learnt is that a rapid rate of progress or too much connectivity comes at a price one may not always want to pay or even know how to deal with.

References

Abbott A (2017) Researchers unite in quest for 'standard model' of the brain. Nature 549:319–321. http://www.nature.com/polopoly_fs/1.22647!/menu/main/topColumns/topLeftColumn/pdf/549319a1.pdf

Aczel A (1999) The Crimean expedition. In: Goldsmith D, Bartusiak (eds) E=Einstein: his life, his thought, and his influence on our culture. Sterling Publishing, New York, pp 81–95. The Aczel article is from God's equation: Einstein, relativity and the expanding universe. Thunder's Mouth Press

Alcoforado F (2015) Planet earth and its limits on use of natural resources. World Resources Forum, 08 July 2015. https://www.wrforum.org/opinion/planet-earth-limits-natural-resources/

Alfred J (2006) Brains and realities. Trafford Publishing, Oxford. http://wk.ixueshu.com/file/78a4cff8ef4f2f19.html

Bera RK (2019a) Synthetic biology, artificial intelligence, and quantum computing. In: Nagpal ML, Boldura O-M, Baltă C, Enany S (eds) Synthetic biology—new interdisciplinary science. IntechOpen. https://doi.org/10.5772/intechopen.83434. https://www.intechopen.com/books/synthetic-biology-new-interdisciplinary-science/synthetic-biology-artificial-intelligence-and-quantum-computing

Bera (2019b) Patenting artificial intelligence inventions. A response to request for comments on patenting artificial intelligence inventions. Submitted to the USPTO on 21 Oct 2019. https://www.uspto.gov/sites/default/files/documents/Rajendra-Bera_RFC-84-FR-448809.pdf

Bera RK (2021a) COVID-19 viewed from a different lens. Social Sciences Research Network, 10 Aug 2021. https://ssrn.com/abstract=3902583 or https://doi.org/10.2139/ssrn.3902583

Bera RK (2021b) Knowledge and employability: the futility of rote education. Social Sciences Research Network, 20 Aug 2021. https://papers.ssrn.com/sol3/papers.cfm?abstract_id=3908420

References

Bera RK, Raj S (2017) Disaster management of human resources. Social Sciences Research Network, 19 May 2017. https://doi.org/10.2139/ssrn.2971188

Bernstein WJ (2004) The birth of plenty: how the prosperity of the modern world was created. McGraw-Hill

Best J (2013) IBM Watson: the inside story of how the Jeopardy-winning supercomputer was born, and what it wants to do next. TechRepublic, 10 Sept 2013. http://www.techrepublic.com/article/ibm-watson-the-inside-story-of-how-the-jeopardy-winning-supercomputer-was-born-and-what-it-wants-to-do-next/

Bloch E (1990) Can the U.S. compete? World Link, Jan–Feb 1990

Britannica (n.d.) Light as electromagnetic radiation. Britannica. https://www.britannica.com/science/light/Light-as-electromagnetic-radiation

Chaitin GJ (2003) Leibniz, information, math and physics. arXiv:math.HO/0306303 v2, 21 June 2003. http://arxiv.org/abs/math/0306303

Cipra B (1999) A prime case of chaos. In: What's happening in the mathematical sciences, vol 4. American Mathematical Society. http://www.ams.org/samplings/math-history/prime-chaos.pdf

Crawford K, Calo R (2016) There is a blind spot in AI research. Nature 538:311–313. http://www.nature.com/polopoly_fs/1.20805!/menu/main/topColumns/topLeftColumn/pdf/538311a.pdf

Cyranoski D (2018) Beijing launches pioneering brain-science centre. Nature, 05 Apr 2018. https://www.nature.com/articles/d41586-018-04122-3

Danos M (1997) Ward-Takahashi identities and Nöther's theorem in quantum field theory. Found. Phys 27(7):995–1009

Dean J, Ng A (2012) Using large-scale brain simulations for machine learning and A.I., 26 June 2012. https://googleblog.blogspot.com/2012/06/using-large-scale-brain-simulations-for.html

Dyson FJ (2007) Why is Maxwell's theory so hard to understand? In: 2nd European conference on antennas and propagation (EuCAP 2007). https://doi.org/10.1049/ic.2007.1146 and http://www.damtp.cam.ac.uk/user/tong/em/dyson.pdf

Einstein A (1905) Zur Elektrodynamik bewegter Körper, in Annalen der Physik. An English translation is available as 'On the electrodynamics of moving bodies' at 17:891. https://www.fourmilab.ch/etexts/einstein/specrel/www/

Elsayed GF, Goodfellow I, Sohl-Dickstein J (2018) Adversarial reprogramming of neural networks. arXiv, arXiv:1806.11146v2 [cs.LG] 29 Nov 2018. https://arxiv.org/pdf/1806.11146.pdf

Erdős P, Rényi A (1960) On the evolution of random graphs. Publ Math Inst Hungar Acad Sci 5(1):17–61. http://citeseerx.ist.psu.edu/viewdoc/download;jsessionid=8C8C9989FF09A1DED54302097A23D951?doi=10.1.1.153.5943&rep=rep1&type=pdf

EurekAlert (2018) Blue brain project releases first-ever digital 3D brain cell atlas. Press Release, 28 Nov 2018. https://www.eurekalert.org/pub_releases/2018-11/f-bbp112118.php

Feynman R (1965) The character of physical law. Modern Library Edition, 1994, Originally published by BBC in 1965, and in paperback by MIT Press, 1967

Feynman RP (2005) The relation of science and religion. In: The pleasure of finding things out. Basic Books, New edition (16 Mar 2005), Chap. 13. http://calteches.library.caltech.edu/49/2/Religion.htm

GenomeWeb (2017) AI for peer review, GenomeWeb, 21 Feb 2017. https://www.genomeweb.com/scan/ai-peer-review

Gibney (2016). Gibney, E. Google masters Go, Nature 529, 28 January 2016, pp. 445–6. http://www.nature.com/polopoly_fs/1.19234!/menu/main/topColumns/topLeftColumn/pdf/529445a.pdf

Goldsmith D, Bartusiak (eds) (2006) E=Einstein: his life, his thought, and his influence on our culture. Sterling Publishing, New York

Gödel K (1931) Über formal unentseheidbare Sätze der Principia Mathematica und verwandter Systeme I. Monatshefte für Mathematik und Physik 38:173–198. (On formally undecidable propositions of Principia Mathematica and related systems I.) (Visit http://jacqkrol.x10.mx/assets/articles/godel-1931.pdf for an English translation by B. Meltzer.)

HBP (n.d.) Human brain project. https://www.humanbrainproject.eu/en/

Headrick DR (1981) The tools of empire: technology and European imperialism in the nineteenth century. Oxford University Press, New York
Heisenberg W (1958) Physics and philosophy: the revolution in modern science. George Allen and Unwin. https://archive.org/stream/PhysicsPhilosophy/Heisenberg-PhysicsPhilosophy#page/n1
Hofstadter D (1979) Gödel, Escher, Bach: an eternal golden braid. Basic Books
Hnilo AA (1985) Chaotic (as the logistic map) laser cavity. Opt Commun, 53(3):194–196. https://www.sciencedirect.com/science/article/abs/pii/003040188590330X?via%3Dihub
Hublin J-J et al (2017) New fossils from Jebel Irhoud, Morocco and the pan-African origin of *Homo sapiens*. Nature 546:289–292. http://www.nature.com/articles/nature22336
IBI (n.d.) About us. https://www.internationalbraininitiative.org/about-us. Accessed 25 Aug 2021
IBI (2020) International brain initiative: an innovative framework for coordinated global brain research efforts. Neuron 105(2):212–216. https://www.cell.com/neuron/pdf/S0896-6273(20)30002-7.pdf and Open Access https://doi.org/10.1016/j.neuron.2020.01.002. Correction at https://www.cell.com/neuron/fulltext/S0896-6273(20)30145-8#relatedArticles
IBL (2017) An international laboratory for systems and computational neuroscience. Neuron 96(6):1213–1218. https://www.cell.com/neuron/fulltext/S0896-6273(17)31136-4 and https://doi.org/10.1016/j.neuron.2017.12.013
Kaku M (2014) The future of the mind. Doubleday
Kurzweil R (1999) The age of spiritual machines: when computers exceed human intelligence. Viking Press
Kurzweil R (2001) The law of accelerating returns, 07 Mar. http://www.kurzweilai.net/the-law-of-accelerating-returns
Kurzweil R (2005) The singularity is near: when humans transcend biology. Viking. http://stargate.inf.elte.hu/~seci/fun/Kurzweil,%20Ray%20-%20Singularity%20Is%20Near,%20The%20%28hardback%20ed%29%20%5Bv1.3%5D.pdf
Landauer R (1991) Information is physical. Phys Today 44(5):23. https://doi.org/10.1063/1.881299. http://www.w2agz.com/Library/Limits%20of%20Computation/Landauer%20Article,%20Physics%20Today%2044,%205,%2023%20(1991).pdf
Mandelbrot BB (1982) The fractal geometry of nature. W. H. Freeman and Co., San Francisco
Markram H (2006) The blue brain project. Nat Rev Neurosci 7:153–160. https://www.nature.com/articles/nrn1848
Matson J (2010) Famed mathematician Benoit Mandelbrot, father of fractal geometry, dead at 85. Scientific American, 18 Oct 2010. https://blogs.scientificamerican.com/observations/famed-mathematician-benoit-mandelbrot-father-of-fractal-geometry-dead-at-85/
Maxwell JC (1873) A treatise on electricity and magnetism, pp x–xi. https://www.aproged.pt/biblioteca/MaxwellI.pdf
May RM (1976) Simple mathematical models with very complicated dynamics. Nature 261:459–467. http://abel.harvard.edu/archive/118r_spring_05/docs/may.pdf
McShane S, Dorrier J (2016) Ray Kurzweil predicts three technologies will define our future. https://singularityhub.com/2016/04/19/ray-kurzweil-predicts-three-technologies-will-define-our-future/#sm.000042uzat13jffbfpj9jucowm0hs
NAP (23640) (2017) Advancing concepts and models for measuring innovation: proceedings of a workshop, NAP, Washington, D.C.. http://www.nap.edu/23640
Nielsen MA, Chuang IL (2000) Quantum computation and quantum information. Cambridge University Press. [Errata at http://www.squint.org/qci/]
NIH (n.d.) What is the brain initiative? National Institutes of Health. https://www.braininitiative.nih.gov/?AspxAutoDetectCookieSupport=1
Nöther E (1918) Invariante Variationsprobleme. (In German). Nachrichten von der Gesellschaft der Wissenschaften zu Göttingen. Mathematisch-Physikalische Klasse, pp 235–257. https://de.wikisource.org/wiki/Invariante_Variationsprobleme English translation by Tavel MA. Nöther E (1971) Invariant variation problems. Transp Theory Stat Phys 1(3):186–207. arXiv:physics/0503066v3 [physics.hist-ph] 30 May 2018. https://arxiv.org/pdf/physics/0503066.pdf

References

OECD (n.d.) Technological and institutional innovation. The World Economy, OECD, (undated). http://www.theworldeconomy.org/advances/advances3.html

Ogle R (2007) Smart world: breakthrough creativity and the new science of ideas. Harvard Business School Press

Popper K (1994) In search of a better world: lectures and essays from thirty years. Routledge, London. Translated by Laura J. Bennett, with additional material by Melitta Mew

Richter D et al (2017) The age of the Hominin fossils from Jebel Irhoud, Morocco, and the origins of the middle stone age. Nature 546:293–296. http://www.nature.com/articles/nature22335

Rubenstein E (2008) Innovation: the rallying cry of the 21st century, Letter from the President of the Academy. The New York Academy of Sciences Magazine, Spring, p 2

Schulze B, Sljoka A, Whiteley W (2014) How does symmetry impact the flexibility of proteins? Philos Trans A Math Phys Eng Sci 372:20120041. https://doi.org/10.1098/rsta.2012.0041

Shannon CE (1937) A symbolic analysis of relay and switching circuits. Master of Science thesis, MIT. http://dspace.mit.edu/bitstream/handle/1721.1/11173/34541425-MIT.pdf?sequence=2

Shannon CE (1938) A symbolic analysis of relay and switching circuits. Trans Am Inst Electr Eng 57:471–495. https://www.cs.virginia.edu/~evans/greatworks/shannon38.pdf (This paper is an abstract of Shannon's Master of Science thesis.)

Shannon CE (1948) A mathematical theory of communication. Reprinted with corrections from The Bell System Technical Journal, 27:379–423, 623–656. https://www.cs.ucf.edu/~dcm/Teaching/COP5611-Spring2012/Shannon48-MathTheoryComm.pdf

Simske S (2019) Meta-analytics: consensus approaches and system patterns for data analysis. Morgan Kaufmann

Solon O (2017) Oh the humanity! Poker computer trounces humans in big step for AI. The Guardian, 30 Jan 2017. https://www.theguardian.com/technology/2017/jan/30/libratus-poker-artificial-intelligence-professional-human-players-competition

Somers J (2017) Is AI riding a one-trick pony? MIT Technology Review, 29 Sept 2017. https://www.technologyreview.com/s/608911/is-ai-riding-a-one-trick-pony/

SNMNH (2018) *Homo sapiens*. Smithsonian National Museum of Natural History. Updated 10 July 2018. http://humanorigins.si.edu/evidence/human-fossils/species/homo-sapiens

Stockton (2017) Stockton N If AI can fix peer review in science, AI can do anything. Wired, 21 February 2017. https://www.wired.com/2017/02/ai-can-solve-peer-review-ai-can-solve-anything/

Tegmark M (2014) Our mathematical universe: my quest for the ultimate nature of reality. Knopf

Toth VT (2003) The principle of gauge invariance, 19 Nov 2003. http://www.vttoth.com/gauge.htm and http://en.wikipedia.org/wiki/Gauge_theory

Turing AM (1936) On computable numbers, with an application to the *Entscheidungsproblem*. Proc London Math Soc S2–42 1):230–265 (1936–1937). https://www.cs.virginia.edu/~robins/Turing_Paper_1936.pdf Correction at: Turing AM (1938) On computable numbers, with an application to the Entscheidungsproblem. A Correction. S2–43(1):544–546. http://www.turingarchive.org/viewer/?id=466&title=02

Warshofsky F (1994) The patent wars. Wiley

Weiß J, Schwietring T (2018) The power of language: a philosophical-sociological reflection. Goethe-Institut. http://www.goethe.de/lhr/prj/mac/msp/en1253450.htm

White House (2016a) Preparing for the future of artificial intelligence. White House, 03 May 2016. https://obamawhitehouse.archives.gov/blog/2016/05/03/preparing-future-artificial-intelligence

White House (2016b) Preparing for the future of artificial intelligence. Executive Office of the President, National Science and Technology Council, Committee on Technology White House, Oct 2016. https://obamawhitehouse.archives.gov/sites/default/files/whitehouse_files/microsites/ostp/NSTC/preparing_for_the_future_of_ai.pdf

White House (2016c) Artificial intelligence, automation, and the economy. executive office of the President of the United States, Dec 2016. https://obamawhitehouse.archives.gov/sites/whitehouse.gov/files/documents/Artificial-Intelligence-Automation-Economy.PDF

Wigner EP (1960) The unreasonable effectiveness of mathematics in the natural sciences. Richard Courant lecture in mathematical sciences delivered at New York University, May 11, 1959; published Communications on Pure and Applied Mathematics 13(1):1–14. http://www.maths.ed.ac.uk/~aar/papers/wigner.pdf

Woodie A (2017) Why deep learning may not be so 'Deep' after all. Datanami, 12 Dec 2017. https://www.datanami.com/2017/12/12/deep-learning-may-not-deep/

WRF (2017) Accelerating the resource revolution. World Resources Forum. Geneva, Switzerland, 24–25 Oct 2017. Meeting report. https://www.wrforum.org/wp-content/uploads/2018/03/MeetingReport_WRF-2017.pdf

Yahoo (2016) 56% Indian employees apprehend poor retirement benefits: survey. Yahoo! Finance, 09 August 2016. https://in.finance.yahoo.com/news/56-indian-employees-apprehend-poor-104805988.html

Yong E (2016) A new origin story for dogs. The Atlantic Daily, 02 June 2016. https://www.theatlantic.com/science/archive/2016/06/the-origin-of-dogs/484976/

Chapter 3
Knowledge and Employability: The Futility of Rote Education

Abstract Our future employability and survivability will depend on our ability to competitively coexist with AI-embedded machines in the job market. We have to be smarter than AI machines. The alternative is working in the gig economy or finding a rare benevolent benefactor. Darwin's theory of evolution says our existence depends on how Nature selectively weeds out the unfit in a given environment. Progressively we have thus arrived at a stage where survival dominantly favors those with superior intelligence and the ability to create new knowledge. At every stage of human evolution—hunter-gatherer, agriculturist, industrialist—survival demanded progressively greater intellectual contributions and competitively productive skills from individuals for success and a dignified place in society. The time has now come when survival will demand even greater intellectual contributions from individuals which rote education cannot provide because it is mechanizable in terms of artificial intelligence. Our future adversaries in the job market will be intelligent machines, other egotistical intelligent *Homo sapiens*, and combinations of them. The heart of AI is algorithmic computation. Computation is all about addition, subtraction, multiplication, division, and comparison of numbers, and problem-solving is all about attaching meaning to numbers.

Keywords Artificial intelligence · Post-industrial economy · Employability · Rationalism

3.1 Our Evolutionary Trajectory[1]

In learning any subject it is important to know what the experts in field do not know; how, why, and what beliefs they hold to be true as a working hypothesis in terms of concepts, ideas, etc. to create a branch of knowledge; and the work culture they adopt to refine, advance, and unify knowledge. We now know that sociology plays a crucial role (some communities and countries are demonstrably head and shoulders above others) in developing knowledge by means of culture and

---

[1] This section is an enlarged version of Bera (2021), Sect. 5.3 Artificial Intelligence (AI).

language. In developing scientific knowledge there is what we call scientific culture, and there is a language we call mathematics (the heart of which is numbers which endows us with the ability to count, measure, and compare, and algorithms (nontrivial mathematical operations) to edit numbers. We also know that any rationally structured language used in communication can be translated into arithmetic whether it be a natural language, image, audio–video script, etc. The present generation takes these for granted when they use digital devices to send emails with text and video attachments, yet much less than a century ago such things were fantasies or possible only in dreams and fairy tales. This chapter explains that doing arithmetic is as simple as taking a walk. If you can walk and count then there is no need to fear artificial intelligence (AI). You can walk your way to creative solutions (admittedly, running is tougher) to complex problems. All you have to do is translate your problem to a number, the desired solution to its own number, and discover the walking route that will take you from one to the other (both backward and forward). Your ability to compete depends on your ability to discover a walking route and attaching meaning to numbers. It is all about finding isomorphisms (see Sect. 3.3.4) and running the fastest. If temperamentally you walk leisurely, you can compensate by being brilliant in discovering isomorphisms, which usually requires a huge dose of talent or serendipity or both.

Arithmetic is a versatile language that can be used for analysis and synthesis. As a naked mannequin it can be used so to say tailor and style clothes (i.e., give meaning through interpretations) for different occasions (i.e., disciplines of knowledge).

The future is bright for those willing to risk treading the untrodden path or discard their addictive dependence on welfare programs and assiduously sharpen their intellectual skills. They must abandon rote education and learn the art of knowledge creation. Our future employability and survivability will depend on our ability to competitively coexist with AI-embedded machines in the job market. The alternative will be the gig economy or a rare benevolent benefactor. In terms of Darwin's theory of evolution our existence depends on how Nature selectively weeds out the unfit in a given environment. We have thus progressively arrived at a stage where survival predominantly favors those with superior intelligence with the ability to create new knowledge. For survival, as hunter-gatherer the *Homo sapiens* sharpened their foraging skills which required negligible intelligence; as agriculturist they shed foraging skills and sharpened their community creation skills through complementary division of labor which required manual dexterity and much higher levels of intelligence for organization and coordination; as industrialists they exchanged manual dexterity for even sharper intellectual skills through rational reasoning and advocating it through rote education. At every stage, survival demanded greater intellectual contribution and competitive productive skills from individuals for success and a dignified place in society. The time has now come when further survival will demand far greater intellectual contributions from individuals and the shedding of rote education, which by its very nature is mechanizable in terms of artificial intelligence. In all this we should not forget that even though Nature is "red in tooth and claw" when it comes to the survival of the species, we are the only species on

3.1 Our Evolutionary Trajectory

Earth which wages wars, massacres, and even engages in senseless brutal killings to decimate itself beyond that demanded by the law of natural selection.

Like beauty for which we have no definition, we know it when we see it, so it is with intelligence. And while our teachers may not have told us, so it is with the numbers "one, two, three. ..." in arithmetic, and a point and a straight line in geometry. We just know them by instinct. They too have no definitions! Notwithstanding, anyone with a normal brain does have enough "intelligence" to grasp arithmetic and geometry! Even more amazingly, we not only created arithmetic and geometry, but also programmed it into computers, and used them to design spacecraft and then to navigate vast distances to land men on the moon and bring them safely back to Earth. The Americans did it in July 1969, a mere 65 years after the Wright brothers (Wilbur and Orville), both bicycle mechanics, made history on 17 December 1903, when they successfully flew in a heavier-than-air flying machine of their own design (without the aid of computers), which in its first flight piloted by Orville, stayed afloat for 12 s and flew a distance of 120 feet and in its fourth and last flight piloted by Wilbur flew for 59 s and traveled 852 feet. Indeed, as Neil Armstrong, the first person to walk on the moon, on 20 July 1969, said on taking his first step on the Lunar surface, "That's one small step for a man, one giant leap for mankind." (See Fig. 3.1.)

Within a half-century of that giant leap, progress in mathematics and computers has primed the world to a state where it is about to leap into a new AI-driven era. In this era survival will demand that we routinely make intellectual leaps to remain employable because AI-embedded machines will be more productive than most rote educated humans. The middle class is about to shrink rapidly.

Fig. 3.1 Aerial leaps. (Left) Wright First Flight 17 December 1903. https://commons.wikimedia.org/wiki/File:Wright_First_Flight_1903Dec17_(full_restore_115).jpg (Public domain) (Right) July 20, 1969: One Giant Leap For Mankind. Images credit NASA. (Top) https://www.bbc.com/news/science-environment-48907836 (Bottom) https://www.space.com/16758-apollo-11-first-moon-landing.html (in public domain)

Table 3.1 Geological timeline

Time	Event	Time	Event
4.6 billion years ago	Origin of the Earth	225 million years ago	The dinosaurs evolve
3.8 billion years ago	First life begins	200 million years ago	The mammals evolve
2.1 billion years ago	Eukaryotes evolved	150 million years ago	First birds
1.1 billion years ago	First sexually reproducing organisms	130 million years ago	Flowering plants evolve
570 million years ago	First arthropods evolve	100 million years ago	The first bees evolve
530 million years ago	The first fish	65 million years ago	Dinosaurs and ammonites become extinct
475 million years ago	First land plants	14 million years ago	The first great apes appear
385 million years ago	First forests	2.5 million years ago	Genus Homo evolves
370 million years ago	The first amphibians	300,000 years ago As estimated in 2017	*Homo sapiens* evolves
320 million years ago	The earliest reptiles	10,000 years ago	End of the last Ice Age

Source Geological Timeline. http://www.bbc.co.uk/nature/history_of_the_earth

> The true test of intelligence is not how much we know how to do, but how we behave when we don't know what to do.[2]

To understand intelligence, we need to understand the creation and evolution of life, especially of the *Homo sapiens*. The origin of life is shrouded in mystery. What we know comes from archeology (the study of human history and prehistory through the excavation of sites and the analysis of artifacts and other physical remains), Darwin's theory of evolution, and most recently, effectively, and reliably through molecular biology and dating methods. Table 3.1 provides a timeline from the origin of the Earth to the end of the last Ice Age. In this timeline, we the *Homo sapiens* have evolved from the Great Apes (they appeared some 14 million years ago). Humans diverged from apes (chimpanzees, specifically) toward the end of the Miocene[3] ~ 9.3–6.5 million years ago. The evolutionary history of apes and humans unearthed so far is largely incomplete.[4]

Among the *Homo sapiens* we are 99.9% genetically identical. Any two humans are genetically vastly more similar than, say, a Western and Central African Chimpanzee. Nevertheless the differences allow scientists to pick out signatures that trace our

---

[2] Holt (1964).

[3] The *Miocene* Epoch, 23.03–5.3 million years ago, was a time of warmer global climates than those in the preceding Oligocene or the following Pliocene.

[4] Almécija (2021).

3.1 Our Evolutionary Trajectory 77

Fig. 3.2 Recent evolutionary history of *Homo sapiens. Source of figure* https://scienceleadership. org/blog/are_humans_still_evolving-2. The original author of this picture seems to be unknown

genetic ancestry and identify startlingly recent instances of human evolution. At a very fundamental level genes define us. We began life as hunter-gatherers and bred according to Darwin's theory of evolution, i.e., the survival of the fittest in a given environment via genetic mutations (see Fig. 3.2). Many adaptive genes increased in frequency during the Neolithic (agricultural) revolution some 12,000 years ago, due to the drastic changes in diet. Compared to the hunter-gatherers who were dominated by the environment and lived at its mercy, the agriculturists and their progenies slowly and gradually learnt to bend the environment so that even the less fit could survive.

The succeeding Industrial Revolution (1760 to 1840) brought about rapid and far reaching developments that allowed us to bend the environment further where even survival of the unfittest became possible by expressions of compassion and empathy by the fittest. The welfare society was born with the oozing milk of human kindness. It encouraged the unfit to become sloths, work-shy, and a growing burden on society. Its tremendous success and excess can be seen in Fig. 3.3a, especially since 1928 when the world's population reached 2 billion. It took us more than 300,000 years to reach one billion in 1803; in 2022, we were at 8 billion and still racing ahead. Pandemics (see Fig. 3.3b) and World Wars[5] have hardly slowed down population growth. Life expectancy has doubled in the last 100 years. The global population is now churning itself to separate the intellectual cream from the rest—a smart, liberal, educated elite that drives and thrives on AI; and the rest seeking employment in a gig economy. In an AI-enabled world as more jobs are automated, the Himalayan problem will be dealing with too many dispensable job seekers. Those with potential to fill good jobs with career paths will need extensive, expensive, and exclusive skills training, coaching, mentoring, and even some hand holding.

People falling into the gig economy from a middle-class perch is a tragedy, more so when they are also grieving from the loss of near and dear ones due to COVID-19. The gig economy shears away personal dignity, their sense of purpose and aim in

---

[5] Hedges (2003). At least 108 million people were killed in wars in the twentieth century. Estimates for the total number killed in wars throughout all of human history range from 150 million to 1 billion. War has several other effects on population, including decreasing the birthrate by taking men away from their wives. The reduced birthrate during World War II is estimated to have caused a population deficit of more than 20 million people.

a
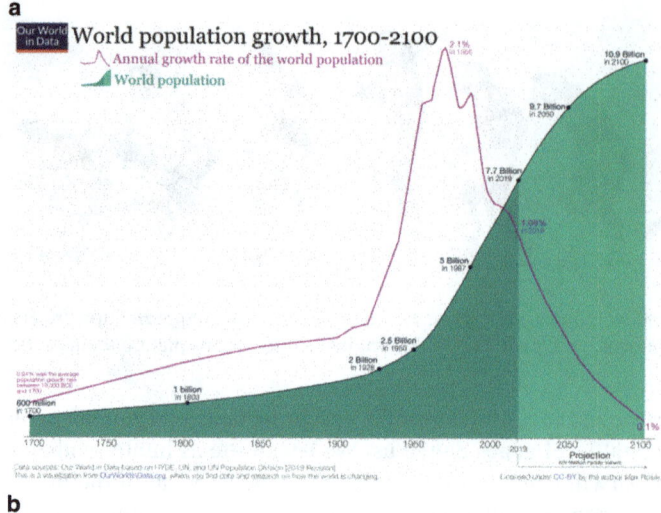

b
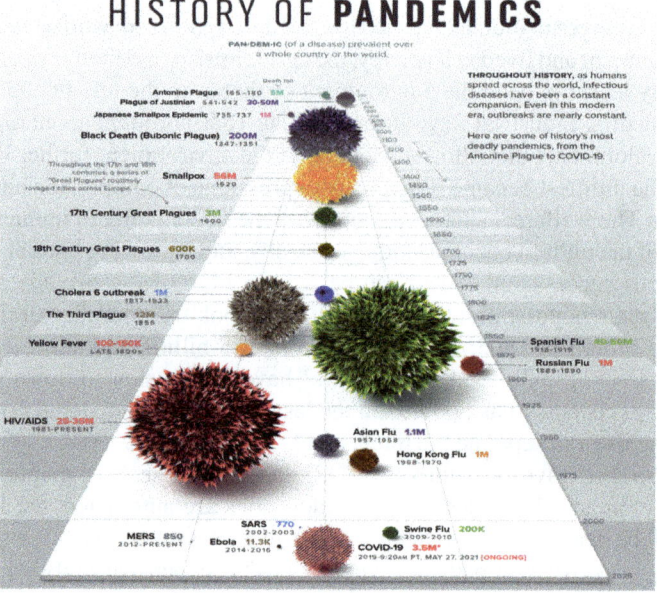

Fig. 3.3 a World population growth from 1700 to 2100. *Source of figure* Our World in data. https://upload.wikimedia.org/wikipedia/commons/5/51/World_population_growth%2C_1700-2100%2C_2019_revision.png. **b** History of pandemics. *Source of figure* LePan (2020). https://www.visualcapitalist.com/history-of-pandemics-deadliest/ "Visualizations are free to share and post in their original form across the web—even for publishers. Please link back to this page and attribute Visual Capitalist"

3.1 Our Evolutionary Trajectory

life, and an unfulfilled craving sets in for the human aspects of work which gave them social and interpersonal connections with their colleagues and superiors and a value-added sense of shared identity in society. Returning to the middle class has suddenly become impossible.

A nagging question arises. Did we defy Nature by proliferating our species by damaging the environment and is COVID-19 Nature's suppressive response to it? One day we may know the answer when we understand global warming better. We bring this up just to provide a quick preview of what climate change involves so that we have some feel for what may happen to us within the next two centuries. Remember that we now see a lot of people walking around in their late 80s and early 90s. So the need to deal with life-and-death upheavals caused by climate change may be just a few decades away.[6]

Global warming has the potential to make us extinct because our survivability is possible only within a narrow temperature band of a few °C. It will hugely affect where we can live, where we can migrate to as immigrants, where we can find employment, and where we can find training opportunities to become employable. Employability will focus on our ability to use our intelligence not just for competing against fellow *Homo sapiens* but against AI-embedded machines. So far the world has survived on the superintelligence of only a few who created and spread rational scientific knowledge,[7] the foundation on which the Industrial Revolution succeeded in mitigating risk to life that Nature constantly thrusts upon us and provided employment to those who could adapt to rote education in a competitive, socio-economic environment.

The upshot of this was that the possibility of building a secure and stable future by the less educated based on inheritance from highly rich parents or royalty began to recede and those of creatively intelligent people (supported by huge numbers of rote educated people in science and technology) began to ascend. The rapid rise of the middle class was propelled by the rapid expansion of science and technology education at the university level and similar rapid rise in the expansion of research laboratories, and the profitable conversion of new knowledge into commercial activity, safeguarded by intellectual property rights, especially in the form of patents. The supply–demand found an equilibrium where knowledge workers were well paid, treated with dignity, and could look forward to progressively enriching career paths that led to middle-class prosperity and a secure retired life. Many acquired wealth exceeding those of the inherited rich.[8] The scale was both impressive and unprecedented. Most of the world's wealthiest people today are not the inherited rich.

The process of industrialization began with select countries—Britain, the USA, Germany—in the West but eventually it percolated to other parts of the world, including China, India, Brazil, etc. mainly through licensed manufacturing and low end IT services that brought substantial income from exports of goods and services rather than by advancing breakthrough technology which continues to be dominated

---

[6] See, e.g., IPCC (2021).
[7] See, e.g., Bernstein (2004).
[8] This is well documented in Piketty (2014).

by the select countries. This has major implications. Countries which could churn out a rote educated workforce on schedule via their university system did create large swaths of aspirational middle class, i.e., they had disposable income and enough money to consume things like fridges or think about buying a car or putting their children into universities who could then join the educated workforce. On the flip side they also became vulnerable to future economic crises. In a global economic crisis the select countries have a much higher survival potential than others; they can cut down on imports, protect their workforce by reskilling programs, and even selectively cut down their exports of raw material and intermediate products to competitors who need them to export finished products. Such actions create an immediate hit on employability of the rote educated middle class of developing and under-developed countries. As the middle class lose jobs, they slide into poverty with the added burden of servicing aspiration-driven debts in the form of Equated Monthly Instalments (EMIs) they owe to banks and financial institutions or face severe penalties for defaulting. The government loses on tax and its ability to support the poor.

The present scenario is the following[9]:

- An overpopulated world where those without rational and useable knowledge are a huge liability which no welfare system can support forever.
- A huge part of the rote educated workforce will lose their jobs to AI-embedded machines and slide into poverty.
- In an uncertain COVID-19 environment, individuals face a two-front war which must be fought by improving one's intellectual and creative skills and staying healthy.
- Those with problem-solving skills at conceptual levels where AI machines are still very weak will be sought after. To produce them the education system must be radically changed, and it can be done only on a small scale at great expense. Our random genetic mutations will ensure that. We not only vary in intelligence, but we are also intellectually lazy even in acquiring rote education.
- People who are currently less vulnerable are competent medical practitioners (sickness will always exist) and honest lawyers (there will always be enough reason to litigate) to the extent their clients can pay for their services. Business will be low for the mediocre.
- The elite class will comprise outstanding teachers and mentors, but they will engage with the most talented and be socially invisible to others.

This means that the industrial era supported by a rote educated workforce is in its terminal phase, and a new era with unusual characteristics will replace it. The transition will be messy because Darwin's law of natural selection will dominate the environment where COVID-19, climate change, and rising artificial intelligence are in concurrent play. Each is a force of immense power over which we have no control. This is not the time to delude ourselves that somehow events will arrange themselves and we will come out unscathed. When Nature seeks a new equilibrium, species become expendable. Our survival will therefore demand not just greater and

---

[9] Bera (2021).

3.2 How Minds Acquire Knowledge

better use of intelligence but also how we integrate with AI systems to blend with the emerging environment.

3.2 How Minds Acquire Knowledge

Let us now see how human minds acquire knowledge and the most fundamental activity an AI machine is involved in. Our current best framework for acquiring knowledge rests on siloed axiomatic (or formal) systems. A grand, unified, single axiomatic system does not yet exist and will never exist. We construct axiomatic systems with the following features.

- We select a set of concepts that we feel are very primitive and agree to accept them without definition. These are the undefined concepts of the system. [Undefined concepts, such as *point* and *straight line* in Euclidean geometry]
- We select some statements concerning these undefined concepts that we feel express such primitive truths about the undefined concepts that we are willing to accept them without proof. These are the axioms of the system. [Axioms]
- Using undefined concepts and axioms we begin the process of defining new concepts in terms of the undefined concepts. [Defined concepts]
- And establishing the truth of new statements about these concepts based on the axioms. [Theorems]

The most famous and ancient example of such a system is Euclidean geometry which we studied in school. When Euclid of Alexandria (325–265 BC) (see Fig. 3.4) codified geometry, he assumed that everyone instinctively knew what was meant by a *point* and a *straight line*. Thus assured he stated 5 axioms which he thought everyone would agree were self-evident, hence not in need of proof. These were:

(1) A straight-line segment can be drawn joining any two points.
(2) Any straight-line segment can be extended indefinitely in a straight line.
(3) Given any straight-line segment, a circle can be drawn having the segment as radius and one point as center.
(4) All right angles are congruent.
(5) If two lines are drawn which intersect a third in such a way that the sum of the inner angles on one side is less than two right angles, then the two lines inevitably must intersect each other on that side if extended far enough. (The unique parallel postulate, i.e., two parallel lines are equidistant.)

What is intriguing about Euclidean geometry is that all the theorems we learnt in school only restate the axioms, i.e., our beliefs. In fact, if we took any five independent theorems (i.e., none of the five are derivable from the other four) and took them as axioms, Euclid's five axioms would become theorems in *our* geometry. For example, the fifth postulate can be replaced by the theorem "the sum of the angles in any plane triangle is equal to two right angles" and the fifth postulate derived as a theorem.

Fig. 3.4 Euclid of Alexandria

The merit of Euclid's axioms is that they are more believable than say the Pythagorean theorem, or that the sum of the interior angles of a triangle will always add to 180°, or the triangle inscribed inside a circle with the diameter as one of its side will be a right-angled triangle, etc. (See Fig. 3.5.) They all look too complicated for our intuition to accept them as true without "proof". Euclidean geometry teaches us how from simple "facts" we can draw complicated conclusions (theorems). This lies at the root of knowledge creation—selecting believable beliefs. Religions cannot be cast as axiomatic systems since the notion of God is not universally believed in or when it is, the belief appears in widely different forms and is often stated in contradictory forms (true in one and false in another). Much water has flown under the bridge since the time when Napoleon Bonaparte (1769–1821) asked Pierre-Simon Laplace (1749–1827) if it was true that there was no mention of the solar system's Creator (i.e., God) in his magnum opus on Celestial Mechanics, Laplace replied, "Je n'avais pas besoin de cette hypothèse-là." (*I had no need of that hypothesis.*)[10] In less than two centuries since, God's standing in the scientific community would irrevocably plunge and with it his power to subjugate humanity by fear and intrigue has declined too.

Since theorems are derivable from the axioms, we do not gain any new information outside of the axioms that we accepted on faith. Of course, for an axiomatic system to work, it must be consistent, i.e., within the system it should not be possible to draw contradictory conclusions from the same input. In an inconsistent system you can draw whatever conclusion you wish to draw.

---

[10] Laplace (1822).

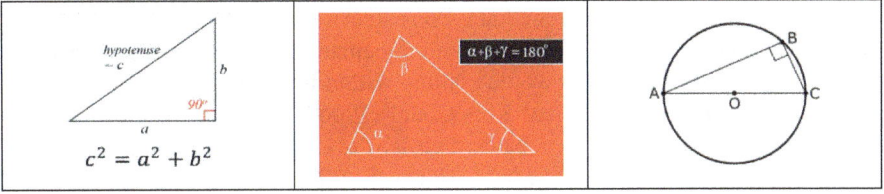

Fig. 3.5 Three randomly chosen theorems from Euclidean geometry

Table 3.2 Isomorphism between formal and computational systems

Formal system	Computational system
Axioms	Program input or initial state
Rules of inference	Program interpreter
Theorem(s)	Program output
Derivation	Computation

Ref: Lewis, J. P., Large limits to software estimation, ACM Software Engineering Notes, Vol. 26, No. 4, July 2001, pp. 54–59

The structure of an axiomatic system begins with a specification (a set of symbols and a grammar or typographical rules for combining the symbols into statements) regarding the construction of syntactically correct statements (well-formed formulas), a set of axioms (propositions regarded as self-evident truths or postulates; they may be any number including zero and infinity), and a finite set of inference rules that allow theorems to be generated using the axioms and previously derived theorems. This is what mathematicians call *first-order predicate calculus*.[11]

Creating an axiomatic system seems to be a non-mathematical and a highly intelligent act. Developing a sequence of theorems with a specific non-trivial objective (i.e., developing algorithms) too seems to be an intelligent act. However, executing a known algorithm is fully mechanizable; it does not require intelligence, in fact, none at all. In mathematics symbols have no meaning other than those implied by their relationships to one another. It therefore lends itself to automation via the isomorphism (i.e., a one-to-one correspondence between two sets of entities) shown in Table 3.2.

During 1900–1905, David Hilbert (1862—1943) was inspired to develop a program to axiomatize all of mathematics (essentially arithmetic). He believed it

---

[11] Predicate calculus: the branch of symbolic logic that deals with propositions containing predicates, names, and quantifiers. A predicate is something which is affirmed or denied concerning an argument of a proposition. A quantifier is an expression (e.g., all, some, greater than, non-negative, etc.) that indicates the scope of a term to which it is attached. First-order predicate calculus uses quantified variables (a variable is a quantity that may change within the context of a mathematical problem or experiment) over non-logical objects and allows the use of sentences that contain variables, so that rather than propositions such as "Alice is a girl" in a natural language, one can bombastically but with greater clarity say, "there exists an x such that x is Alice and x is a girl", where "there exists" is a quantifier, and x is a variable. This distinguishes it from propositional logic, which does not use quantifiers or relations. In this sense, propositional logic is the foundation of first-order predicate calculus (also called first-order logic).

was possible to find a belief system that is consistent and complete.[12] The aim was to discover "What constitutes a valid proof in mathematics and how is such a proof to be recognized?" Such a system would contain within itself an algorithm for testing the validity of proofs. A skeptical Kurt Gödel (1906—1978) in 1930 proved that such a system was impossible.[13] Nevertheless, the creation of axiomatic systems is an amazingly "intelligent" act ever performed by *Homo sapiens* since they first appeared on Earth. Equally amazing is the discovery about a century ago that any axiomatic system can be arithmetized, that is, put into an arithmetical and computable form and hence implemented on a Universal Turing Machine (UTM) (see Sect. 3.3).

3.3 Universal Turing Machine

Alan Turing began by defining a simple abstract mathematical model of a human computer who robotically (and thus mindlessly) executes mathematical operations as ordered without the benefit of insight, foresight, or intuition. He then showed that it could simulate any classical computation that humans could perform. In short, classical computation was completely mechanizable. His mathematical model is now famously known as the Universal Turing Machine (UTM). The UTM comprises a small central processing unit (CPU) with a memory of only a few bits—what is technically known as a 'finite-state automaton,' since it only has a finite number of different internal memory states (see Fig. 3.6). The CPU can read or write binary digits on an infinitely long computer tape as it moves back and forth according to its own very simple, hard-wired rules. This single tape serves as the Turing machine's input, program, bulk memory, output, and even 'garbage' storage. It describes all classical computers we use today. See Fig. 3.6 for a quick description. See also Sects. 3.3.2 and 3.3.3.

The Universal Turing Machine (UTM) raised the tantalizing possibility that, in principle, discovering axiomatic systems itself may be mechanized if machines could be programmed to introspect, i.e., programmed to execute a certain class of iterative (self-referential) functions.

It is this aspect of mechanized knowledge creation that I and a small group of researchers working with me are engaged in. The inspiration for this research came from the pioneering work of Noam Chomsky who in the 1950s defined a language

---

[12] Hilbert announced the initial version of his project in a series of papers and lectures between 1900 and 1905. However, Henri Poincaré had serious doubts about the feasibility of such a project (see Science et méthode. Paris, 1908, 311 pp. Volume II, Chapter IV). He pointed out that in trying to prove the consistency of mathematics by means of the induction principle Hilbert would use a circular argument: the consistency of mathematics therefore implies the consistency of the induction principle proved by means of the induction principle! Few people at the time realized the deep significance of this remark. The future would prove Poincaré right! For a historical perspective, see Grattan-Guinness (2000).

[13] This epoch-making discovery of Kurt Gödel, a young Austrian mathematician (at age 24), was announced by him to the Vienna Academy of Sciences in 1930 and later published in Gödel (1931).

3.3 Universal Turing Machine

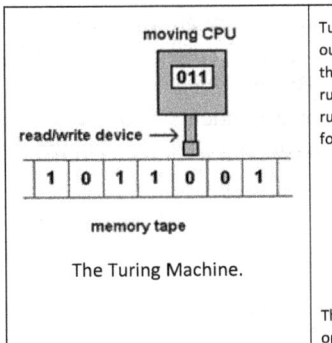

moving CPU [011] read/write device → | 1	0	1	1	0	0	1 | memory tape The Turing Machine.	Turing's approach is as follows. Let f be a function that takes a bit-string and outputs a bit. Then an algorithm for computing f is a set of mechanical rules, such that by following them we can compute $f(x)$ given any input $x \in \{0, 1\}$. The set of rules being followed is finite and must work for all infinitely many inputs; each rule may be applied arbitrarily many times. Each rule must involve one of the following elementary operations: • Read a bit of the input. • Read a bit from the "scratch pad". • Write a bit to the scratch pad. • Stop and output either 0 or 1. • Decide which of the above operations to apply based on the values that were just read. The running time of the algorithm is proportional to the number of elementary operations performed.

Fig. 3.6 Turing machine. *Source of figure* Benjamin Schumacher. Entropy, complexity, and computation. Natural science colloquium. Kenyon College—September 28, 1989. http://physics.kenyon.edu/coolphys/thrmcmp/newcomp.htm

as the set of "strings" that can be formed by chaining together a group of symbols, whether the symbols are the letters of the English alphabet, the 0s and 1s of computer bits, or whatever. The idea of string manipulation is central to building axiomatic systems and all programming languages meant for execution on a Universal Turing Machine.

3.3.1 Binary Strings

A binary string is a linear sequence of arbitrary length constructed using an alphabet of only two symbols, usually chosen by mathematicians and computer scientists to be 0 and 1. A physical representation of such a string is usually a string of objects where each object (called a bit) can be held in two switchable states, one called 0 and the other 1 according to an agreed upon convention. Groups of bits can represent additional symbols (either by convention or on the fly). Binary notation is similar to decimal notation but uses a different base. Decimal numbers use 10 as their base, which means that each digit counts for ten times as much in its present position than had the digit been placed to its right. Binary notation uses base 2, which means that each position counts for twice as much. When it is important to distinguish the base, the text uses a small subscript, like this

$$00101010_2 = 42_{10},$$

where the binary representation 00,101,010 is equivalent to the number 42 in decimal.

A binary string of n bits can display 2^n unique binary configurations. Converting a number from one base to another does not change the number. Numbers do not have bases; their representations do. Sequences of bits by themselves have no intrinsic meaning except for the representation, activity, or meaning they are assigned in an interpretive context.[14] For example, when computing, it may represent data, arithmetic and logical operations, index, messages, etc. By choosing an appropriate grammar for constructing bit strings and providing correspondence rules for their interpretation or assigning meaning (say, via table look-up, e.g., Famous people as in Table 3.3), one can use bits to represent anything one wants—characters using numeric character codes; real numbers using floating-point approximations; two-dimensional bit arrays to represent images; sequences of images to represent video; etc.

In Table 3.3, for example, a reference to the decimal number 6 given a context would mean John Milton if the person's name was sought, or his image if his photograph was sought or a short description of him if that was sought.

3.3.2 What is the Central Activity of a UTM?

So what does a computer do? Well, it goes from one integer to another. Each integer has encoded within it a prescription for generating the next integer! So, Leopold Kronecker (1823–1891) was not off the mark when he said, "God made the integers, all the rest is the work of man."[15] A binary sequence, S, in a digital computer's memory at any instant also codes for an integer, n. In a running computer, after every clock cycle, some of the binary bits will flip and hence change (S, n) say, from (S_1, n_1) to (S_2, n_2). Thus, the computer goes from the integer, n_1 to integer, n_2, and the instructions that bring about the change were obviously encoded in S_1, while S_2 is similarly encoded for the next cycle. Thus, if Nature is isomorphic to a mathematical structure, one may conclude the universe (Fig. 3.7) is indeed a number crunching computer, and hence deeply connected no matter how diverse the universe may look. The universe is going from one number to another.

In an axiomatic system we can pull out a high-level (coarse, chunked, broad) description of the system from its low-level (more detailed, atomic) description by describing the high-level description in terms of theorems. It is at the higher levels that the goals of the system become more visible to a human mind. Humans understand software written in a high-level programming language more easily than its equivalent binary string. We trust a theorem once we understand its proof. Finally, we note that an axiomatic system does mathematical work when its rules are invoked, not otherwise. This aspect is related to the laws of thermodynamics in physics.[16]

---

[14] For an elementary exposure to binary strings, visit Binary Representation and Strings https://web.stanford.edu/class/cs208e/cgi-bin/main.cgi/static/lectures/05-BinaryRepresentation/05-BinaryRepresentation.pdf.

[15] He said it in German: "Die ganzen Zahlen hat der liebe Gott gemacht, alles andere ist Menschenwerk".

[16] For a deeper understanding of this statement, see Landauer (1961).

3.3 Universal Turing Machine

Table 3.3 Short biographies of famous people

Binary code	Decimal number	Person	About the person	Photograph
000	0	Isaac Newton	(1642–1727). Widely recognized as one of the greatest mathematicians and most influential scientists of all time. His book Philosophiæ Naturalis Principia Mathematica (Mathematical Principles of Natural Philosophy) first published in 1687, established classical mechanics Photo: https://en.wikipedia.org/wiki/Isaac_Newton	
001	1	James Clerk Maxwell	(1831–1879). He formulated the classical theory of electromagnetic radiation, unifying for the first time electricity, magnetism, and light as different manifestations of the same phenomenon Photo: https://en.wikipedia.org/wiki/James_Clerk_Maxwell	
010	2	Albert Einstein	(1879–1955). Widely acknowledged as one of the greatest physicists of all time. Einstein is known for developing the theory of relativity. He also made important contributions to the development of quantum mechanics Photo: https://en.wikipedia.org/wiki/Isaac_Newton	
011	3	David Hilbert	(1862–1943). One of the most influential mathematicians of the 19th and early twentieth centuries. Hilbert discovered and developed a broad range of fundamental ideas in many areas of mathematics. In 1900, he presented a collection of problems that set the course for much of 20th-century research in mathematics Photo: https://en.wikipedia.org/wiki/David_Hilbert	

(continued)

Table 3.3 (continued)

Binary code	Decimal number	Person	About the person	Photograph
100	4	Srinivasa Ramanujan	(1887–1920). Indian mathematician who had almost no formal training in pure mathematics, yet he made substantial contributions to mathematical analysis, number theory, infinite series, and continued fractions, including solutions to mathematical problems then considered unsolvable Photo: https://en.wikipedia.org/wiki/Srinivasa_Ramanujan	
101	5	Carl Friedrich Gauss	(1777–1855). German mathematician and physicist who made remarkable contributions in mathematics and science. Sometimes referred to as the *Princeps mathematicorum* ("the foremost of mathematicians") and "the greatest mathematician since antiquity" Photo: https://en.wikipedia.org/wiki/Carl_Friedrich_Gauss	
110	6	John Milton	(1608–1674). English poet and intellectual. He is best known for his epic poem Paradise Lost (1667). Written in blank verse, it is widely considered to be one of the greatest works of literature ever written Photo: https://en.wikipedia.org/wiki/John_Milton	
111	7	Wolfgang Amadeus Mozart	(1756–1791). Mozart showed prodigious ability from his earliest childhood. Though he died young, he composed more than 600 works of virtually every genre of his time. Many of his works are considered pinnacles of the symphonic, concertante, chamber, operatic, and choral repertoire. He is considered among the greatest classical composers of all time Photo: https://en.wikipedia.org/wiki/Wolfgang_Amadeus_Mozart	

Example for table look-up

3.3 Universal Turing Machine

In this Universe the encoded integer stands between you and your employability.

Fig. 3.7 Digital universe. Credit Image: Messier 101, Pinwheel Galaxy. European Space Agency and NASA. Release date 28 February 2006, 12:00. https://esahubble.org/images/heic0602a/

Most axioms and theorems (analogously beliefs and derived customs) humans absorb come from the environment or previous generations or a revered authority. Most humans readily adapt, acclimatize, and conform to their community's beliefs if it helps them in their fight for survival by gaining community support. This herd mentality comes from its gambling instinct when dealing with risk to life. Most human brains and minds are imitative not creative. Thus, when a person embraces a belief system by relinquishing its ability to reason independently it becomes a layperson in the belief system. Such a person, by definition lacking creativity, will fail to compete against AI machines and become trapped in the gig economy.

3.3.3 Capturing Concepts Typographically

Homo sapiens are born with the instinctive ability to capture concepts by simple typographical means. An example is the character string "2 + 3 = 5" (effortlessly recognized by every reader of this book) which captures the concept of numbers in the symbols '2', '3', '5', the addition operation in the ' + ' symbol, and the concept of equivalence in the ' = ' symbol. In 1936, Alan Turing brilliantly used this human instinct to capture concepts by typographical means in his theoretical design of the UTM. The typographical UTM performs the following trivial operations[17]:

1. Reads and recognizes symbols from a finite set of symbols.[18]
2. Writes down any symbol belonging to that set.
3. Copies symbols from the set from one location to another.
4. Erases written down symbols.
5. Compares symbols.
6. Stores and uses a list of previously generated symbol strings.

---

[17] Hofstadter (1979), p. 64.

[18] One can even choose one symbol and use its orientation or color to create an array of uniquely identifiable set of symbols.

Recall that the UTM, conceived before there were any real computers, is not a physical object but a piece of esoteric abstract mathematics. Turing regarded the human brain to be a 'machine' at least when it was engaged in computing. That, he felt, was nothing but calculations performed by a human mathematician who has unlimited time and energy, an unlimited supply of paper and pencils, perfect concentration, and works according to some algorithmic[19] or 'rule of thumb' method (i.e., rote education). (In the context of his seminal paper,[20] the words 'computer', 'computable', and 'computation' pertain to *human calculators*.) Thus, by such simple operations as noted above, the UTM can compute by algorithmic means, whatever is thrown at it packaged as a proper program, e.g., forecast the weather, design atomic weapons, design and fly spacecraft, perform robotic surgery, take-off and land airplanes in heavy fog, explore the subatomic world and the universe at large, etc. By conceptualizing and manipulating numbers (and attaching meaning to them and their stylized and ornamented groupings produced by various operations, e.g., addition, subtraction, multiplication, division, etc.), mathematicians can create an abstract world in which they can mentally visualize the real or any imagined world by typographical means. One such is the elementary arithmetical scheme dealing with the numbers 0, 1, 2, 3, 4, ... (the symbol '...' means that you can extend the sequence as far as you want) that you learnt in primary school.

Primary School Arithmetic

Table 3.4 provides a summary of the arithmetic you learnt in primary school. Let the symbols m, n called operands stand as placeholders for non-negative numbers[21] 0, 1, 2, 3, 4, ... etc., and the symbols $+$, $-$, $\times$, $/$, $=$, $\neq$, $>$, and $<$, respectively, denote operators we call addition, subtraction, multiplication, division, equivalence, non-equivalence, greater than, and smaller than, which can be applied on the operands m, n. Each operand can independently accept any non-negative number. The operators are the only means of arranging, rearranging, and comparing numbers within a permitted string formatted (the way in which something is arranged or set out) in the prescribed style.

An important reminder. In mathematics symbols have no meaning other than those implied by their relationships to one another. Mathematics contains only rules for arranging and rearranging symbols. The preferred arrangement is a linear sequence of symbols (hence called a string). A mathematical structure is like a mannequin on which we test the consistency and completeness of the knowledge we create

---

[19] *Algorithm.* A step-by-step problem-solving procedure, often an established, recursive computational procedure for solving a problem in a finite number of steps.

[20] Turing (1936).

[21] The notion of negative numbers is excluded here. We shall deal with them in the next subsection. "It was not until the nineteenth century when British mathematicians like De Morgan, Peacock, and others, began to investigate the 'laws of arithmetic' in terms of logical definitions that the problem of negative numbers was finally sorted out." – Leo Rogers. The History of Negative Numbers. NRICH, University of Cambridge, 2009. https://nrich.maths.org/5961.

3.3 Universal Turing Machine

Table 3.4 Primary school arithmetic

Symbol, name of symbol	Formatting style and/or meaning	Special cases
$+$, addition	$m + n = n + m$	$0 + n = n + 0 = n$; $m + 0 = 0 + m = m$.
$-$, subtraction	$m - n$ if $m > n$; else $n - m$ if $n > m$.	If $m = n$ then $m - n = n - m = 0$.

$\times$, multiplication	$m \times n = n \times m$	• If m or n or both are 0,
		  then $m \times n = n \times m = 0$.
		• $m \times 1 = 1 \times m = m$;
		  $n \times 1 = 1 \times n = n$;
		  and $1 \times 1 = 1$.

/, division	m / n
	$m / n \neq n / m$ (in general)
		• m / n is invalid
		  if either $n = 0$ or $m = n = 0$.
		• $m / 1 = m$; $n / 1 = n$; and $1 / 1 = 1$.
		• If m / n is valid and $m = n$,
		  then $m / n = n / m = 1$.

$=$, equivalence	The string to the left and the right of the ' $=$ ' sign is interchangeable.	
$\neq$, non-equivalence	The string to the left and the right of the ' $\neq$ ' sign is not interchangeable.	
$>$, greater than; $<$, smaller than	$m > n$ also means $n < m$; $m < n$ also means $n > m$.	
!, negation	!m puts a negative sign before m.	Not taught in primary school.

or of existing knowledge for wear and tear. Mathematics is not an "-ism". The "-ism" comes from the tailor along with his/her preferences, prejudices, tastes, trends, fashions, etc. These aspects often show up during fitting trials.

College-level arithmetic

We now come to the college level arithmetic without which you cannot survive in the AI world. At this level you must know not only about negative numbers but also about many other arithmetic operators and symbols (which we will not discuss in this book) without which you will end up in the gig economy. So let us begin with an extended version of the non-negative numbers, the signed numbers which are either positive, negative, or both.

Let us, therefore, consider the set S of signed integers:

$$S \equiv \{\ldots, -5, -4, -3, -2, -1, \pm 0, +1, +2, +3, +4, +5, \ldots\}$$

Notice the left-right symmetry of the unsigned numbers 1, 2, 3, etc. about the symbol ± 0. Numbers to its left (when you face the page) carry a negative '$-$' sign, those to its right carry a positive '$+$' sign. The number 0 carries both positive and negative signs to show its neutrality to direction and this uniqueness allows each signed number in the sequence to be placed at a specific position with respect to ± 0. The symbol '...' means that you can extend the sequence indefinitely in the direction the numbers are progressing. Take a quick glance at Table 3.5 to familiarize yourself with the symbols we will now use.

92   3 Knowledge and Employability: The Futility of Rote Education

Table 3.5 College-level arithmetic

In the columns below, let the symbols m, n called operands stand as placeholders for the signed numbers in $S \equiv \{\ldots, -5, -4, -3, -2, -1, \pm 0, +1, +2, +3, +4, +5, \ldots\}$

Symbols if not defined or redefined in this table carry the same meaning as in Table 3.4. The exceptions to note are: (1) the '+' and '−' signs are now attached to a number. As we shall see, this eliminates the need to separately define addition and subtraction. They can be subsumed in the concatenation operator. The arithmetic operations will require us to walk back and forth along the string S where each signed number is separated by a ',' from its immediate neighbors. In all cases, the sign of the result will be the sign of the square where your walk ends. Your walk ends at R1 or R2 or ... or R7 as the case may be and denotes the result of the operation carried out, i.e., where you finally find yourself standing in S. Blank spaces appearing in a string are for visual clarity

Symbol, name of symbol	String representation, Formatting style and/or meaning in Table 3.5 notation	String representation, Formatting style and/or meaning in Table 3.4 notation
No symbol, concatenation	$R1 = +m+n = +n+m;$ $R2 = +m-n = -n+m;$ $R3 = -m-n = -n-m.$	$R1 = m+n;$ $R2 = m-n$ if $m > n$ else $!R2 = n-m$ if $n > m;$ $R3 = !R1.$
*, multiplication	$R4 = +m*+n = +n*+m = -m*-n = -n*-m;$ $R5 = +m*-n = -n*+m = -m*+n = +n*-m.$	$R4 = m \times n;$ $R5 = !R4.$
÷, division	$R6 = +m \div +n = -m \div -n;$ $R7 = +m \div -n = -m \div +n.$	$R6 = m/n;$ $R7 = !R6.$
=, equivalence	The string to the left and the right of the '=' sign is interchangeable.	
≠, non-equivalence	The string to the left and the right of the '≠' sign is not interchangeable.	
>, greater than; <, smaller than	$m > n$ also means $n < m;$ $m < n$ also means $n > m.$	
!, negation	!m changes the sign of the number m.	

An unsigned number represents the magnitude of a signed number. For example, −5 and +5 have the same magnitude 5. When we talk of unsigned numbers (i.e., their magnitude only), we can freely use Table 3.4 to deal with them. Within the set S as typographically presented inside the curly brackets in Table 3.5, the left neighbor of any signed number is its predecessor and the right neighbor is its successor. Further, for every signed number, say, $+n$ there is an oppositely signed partner $-n$ so that if they are placed side-by-side as $+n-n$ or $-n+n$ or can be so placed using permitted rules of movement in a string of symbols, they will annihilate each other and vanish or ephemerally exist as a shadow in the form ±0. Further, the signed numbers are notionally separated by a signed "distance" from each other. The distance is signed '+' if our eye movement from a signed number to another signed number is to the right; it is signed '−' if such movement is to the left. The unsigned distance is the magnitude of the signed distance. The unsigned distance between any two neighbouring signed number is 1 stride and its sign is decided by the "from-to" eye movement. Thus the "from-to" signed distance between −3 (from) and +5 (to) is +8 strides and between +5 (from) and −3 (to) is −8 strides. For convenience, if a single movement comprises multiple strides, say, $+n$ strides, we call it a leap of $+n$ strides or $+n$ leap. A 0 stride or ± 0 stride means standing still.

We can now draw an analogy between the nomenclature of walking and arithmetical operations (e.g., addition, subtraction, multiplication, division), and get an instinctive feel for arithmetic. When we walk, we take strides. In your mind think of '+' and '−' as denoting opposite directions. Thus, a +1 stride means going to the successor from one's present location, and likewise a −1 stride means going to the predecessor. A $+n$ stride means taking n contiguous +1 strides in the +ve direction, and a $-n$ stride means taking n contiguous −1 strides in the − ve direction.

Walking the arithmetic

In our walking analogy, when we stand at +6, it means we are +6 strides away from ± 0, and the succeeding signed number is +6+1 = +7, i.e., at +7 distance from ± 0; likewise the preceding number is +6−1 = +5 distance from ± 0. Also, $+n-n = -n+n = \pm 0$ i.e., going back and forth the same distance gets us nowhere!

Addition, subtraction by Concatenation

Now consider the concatenated string −7+4 made from −7 and +4. This means the following walk: Stand at ± 0 and take (or leap) −7 strides and then take (or leap) +4 strides. We will land at −3, i.e., at the expected result. Alternatively, according to Table 3.5, we could have rewritten −7+4 as +4−7 and taken +4 strides and then taken −7 strides, and we still would have arrived at −3. Concatenation permits us to combine addition and subtraction of signed numbers in the same arithmetic operation. In fact, we can do this back and forth walking with any number of signed numbers typographically arranged in a line, e.g., if we walk +10+6+1−8+3−1 we will stop at +11, and this would be true no matter how the numbers +10,+6,+1,−8,+3,−1 are arranged in the string. For example, +10+1−8+3−1+6 will also lead us to +11. This property is called commutative.

Walking through addition and subtraction in arithmetic is a cakewalk! You can think of the linearly, equispaced (in stride units), signed integers as symbols placed on a tailor's measuring tape laid out on a flat surface and walk along the tape. By instinct, we can now relate walking to keeping track of bank accounts by relabeling stride, say, to dollars and '+' and '−' representing credit and debit, respectively. The "intelligent" act here is that we have attached meaning to numbers in a counting context purely by instinct (e.g., +5 is greater than +2), and this instinct can be seen even in babies (e.g., a heap of five identical things is bigger than a heap of two of those things) just as we distinguish a larger line from a smaller line. Such instincts appear to be universally shared among *Homo sapiens* and hard-wired in their brains. However, in extremely rare cases, some individuals are "superwired" in certain rare instincts. Such people are likely to be geniuses.

The unsigned "distance" between any two signed numerals is the smallest number of strides needed to step out from one signed numeral to walk to the other. The unsigned distance between a signed integer and its successor or its predecessor is always 1.

Multiplication

When we multiply two signed numbers, e.g., +5 and −3, the string we write is either +5 * −3 or −3 * +5 (see Table 3.5). Both mean the same thing because they produce the same result. Since the two operands in multiplication are allowed to swap their places in the string, multiplication too is commutative. The multiplication walk essentially means starting at ± 0, ignoring the sign of the operands and taking $+n$ leaps each of $+m$ strides or alternatively, taking $+m$ leaps each of $+n$ strides. Now look at the signs of the two operands. If they are the same, we are already standing at the answer $+a$. If the signs are different, then taking two $-a$ leaps from $+a$ will land us at the result $-a$.

Division

When we divide one signed number (called dividend or first operand of magnitude m) by another signed number (called divisor or second operand of magnitude n), e.g., +5 by −3, the string we write is +5 ÷ −3. (See Table 3.5.) The division walk essentially means starting at ± 0, ignoring the sign of the operands and taking a $+m$ leap (for the first operand), then taking $-n$ leaps (related to the second operand) q times (i.e., as many times as possible without going past ±0; if no $-n$ leap is possible then $q = 0$; q is called the unsigned quotient and must be counted and kept in memory) and stop. If additional strides, say, $-r$ strides remain to reach ± 0, then the unsigned remainder is r, else $r = 0$ (i.e., we are already standing at ±0. Now look at the signs of the two operands. If they are the same, we are already standing at the remainder $+r$ or ± 0. If the signs are different, then taking two $-r$ leaps from $+r$ will land us at the remainder $-r$. In either case q will inherit the sign of r.

(The job of the unsigned divisor is to subtract out as many multiples of itself from the unsigned dividend as it can till nothing is left or what remains is smaller than the unsigned divisor.) If $m = q \times n$ then $r = 0$, else $r = m - q \times n$. If $n > m$, then obviously $q = 0$ and $r = m$.

What did we learn?

Addition, subtraction, multiplication, division, and comparison are the basic operations on which the entire edifice of mathematics is built. They relate to our basic instincts of aggregation, segregation, and comparison by size according to some agreed upon standard of measurement.

Most importantly, if we can add, subtract, multiply, and divide by walking, then surely arithmetic is mechanizable! The only thing you need to know is how to count (Nature more or less teaches that) and change directions forward and backward. Arithmetic is an artifact we created out of basic human instinct—the ability to quantify, to do and undo, to create and erase, and generally to distinguish between opposites. When positive and negative numbers meet (concatenate), they augment their own kind ('+'s get together and '−'s get together), but when they meet opposite kinds they annihilate each other.

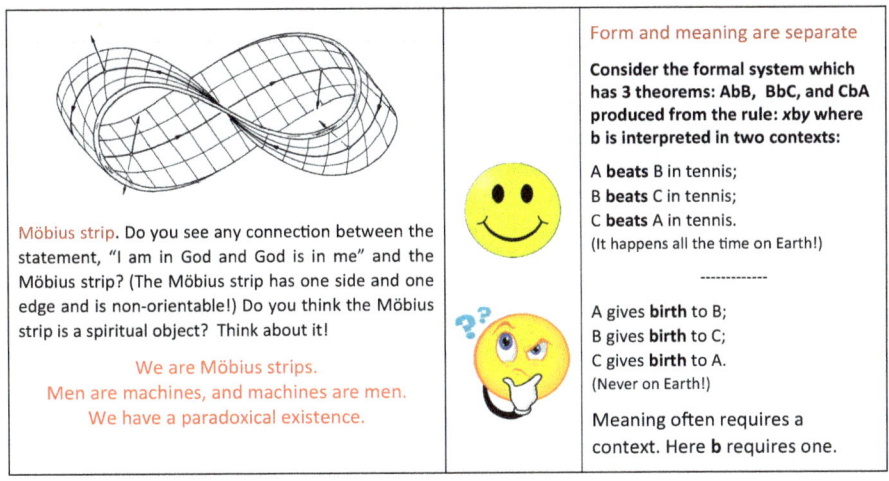

Fig. 3.8 Form and meaning. *Source of figure* (left): Xichen Sheng. 1963. Möbius strip. 21 February 2017. https://medium.com/designscience/1963-88a359d2f68b (Assumed fair use)

3.3.4 Form and Meaning

Form (or shape or template) essentially describes a formatting scaffolding on which information can be placed and stored in prearranged placeholders in an accessible and locatable form. Giving meaning to or interpreting the information populated in a form is an intelligent act especially when multiple, ambiguous, and even contradictory interpretations become possible. See Fig. 2.1 in Chap. 2 for two striking examples.

Sometimes a mere twist in a form can dramatically change the interpretations we can make. Figure 3.8 (left) is such an example. It is a piece of ribbon whose two ends have been brought together after giving one of the ends a half-twist to form what is called the Möbius strip.[22] Defying all our instinctive understanding of the universe, this object has only one side and one edge! Its surface is non-orientable. It has an even more interesting property—it remains in one piece when split down the middle! Does the Möbius strip carry a spiritual meaning: *I am in God and God is in me*? Fig. 2.1 in Chap. 2 and Fig. 3.8 (right) below also imply that form and meaning are separate. Meaning often requires a context. Why bring up the Möbius strip? Because human society is undergoing a half-twist where the distinction between man and machine is being blurred—*Men are machines, and machines are men*. In an AI-dominated world, we have a paradoxical existence.

Let me amplify some more on isomorphism by which we discover meaning or relationship between objects, ideas, concepts, branches of knowledge, etc.

---

[22] It was co-discovered independently by the German mathematicians August Ferdinand Möbius and Johann Benedict Listing in 1858.

Isomorphism

We first introduced isomorphism in Sect. 3.2, Table 3.2 where an analogy was established between formal and computational systems. Engineers are deeply familiar with another example of isomorphism. In our known universe, two different physical systems obeying different laws of physics, e.g., one mechanical and the other electrical, are isomorphic to each other because they share a common mathematical model. This means that any member in the trio (mathematical, electrical, mechanical) can simulate the other two. Further, by virtue of Table 3.2 each can be simulated on a UTM. An AI machine need only know the mathematical model. To meaningfully relate to a world, it only needs a look-up table for the relevant isomorphism. Consider, e.g., a spring-mass-damper system in mechanical engineering and an RLC circuit in electrical engineering as shown in Fig. 3.9. The mathematical description of the two systems has the same mathematical form, which then allows us to establish the isomorphism that exists between them.

Formally, the word isomorphism applies when two complex structures can be mapped onto each other in such a way that to each part of one structure there is a corresponding part in the other structure, where *corresponding* means that the two parts play similar roles in their respective structures. In this sense, it is an information-preserving mapping. Since isomorphism may come in various forms, it is very likely that one may miss it when encountered. Hence,

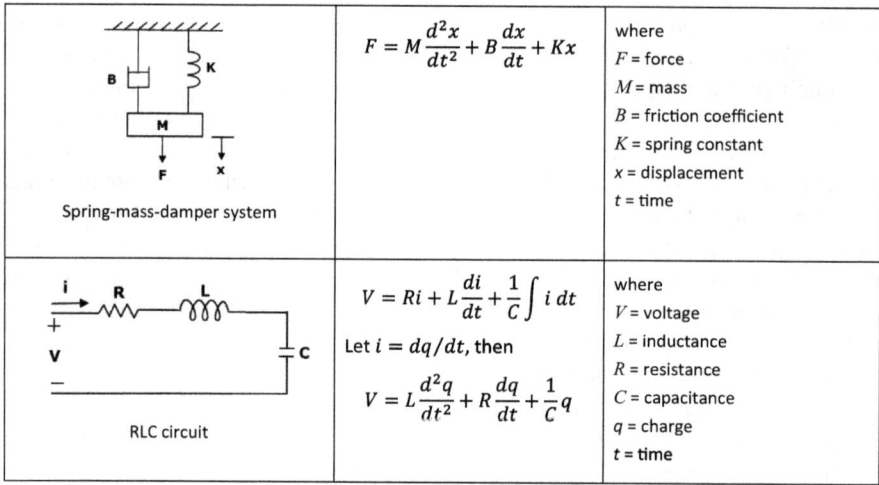

Fig. 3.9 Isomorphism between electrical and mechanical systems. By comparing the entries in the two rows in the third column, the isomorphic pairings (force, voltage), (mass, inductance), (friction coefficient, resistance), (spring constant, reciprocal of capacitance), (displacement, charge), and (time, time) become obvious

3.3 Universal Turing Machine

Table 3.6 Isomorphism between axiomatic (formal) and other systems

Formal system	Computational system	Judicial system	Biological system	Governance system	Natural system
Axioms	Computer	Constitution	The DNA	Government	Nature
Rules of inference	Program interpreter	Judicial interpretation	Laws of biochemistry	Procedural rules	Laws of Nature
Theorem(s)	Program output	Judicial judgment	Biomolecules	Administrative decisions	Output based on input
Derivation	Computation	Legal arguments	Biochemical reactions	Executive action	Derivation, computation

> It is a cause for joy when a mathematician discovers an isomorphism between two structures which he knows. It is often a "bolt from the blue", and a source of wonderment. The perception of an isomorphism between two known structures is a significant advance in knowledge – and I claim that it is such perceptions of isomorphisms which create *meanings* in the minds of people.[23] — Douglas Hofstadter

Thus, by convention, when we find an isomorphic relationship between an axiomatic (formal) system in mathematics and another system (abstract or physical), we say we have given the axiomatic system a meaning (or an interpretation or established an equivalence) in the nomenclature of the other system. See Table 3.6 for examples. Isomorphism between systems is always reciprocal, that is, one system gives meaning to the other. It essentially means that one system can simulate the other in their respective universe. It also means that if two systems are isomorphic, their formal mathematical representation will be identical. In this sense, isomorphism is an information-preserving mapping, i.e., the same information can be represented by either system completely and accurately without any loss or gain.

Isomorphism thus means two things. (1) The information content of any axiomatic system can be translated into arithmetical form without any distortion and vice versa; and (2) therefore any axiomatic system can be simulated on a Universal Turing Machine (UTM). This means information processing and computing are two faces of the same coin. This became exquisitely clear when Claude Shannon presented his information theory in 1948.[24] Shannon's theory makes AI possible. Rote education too often ignores this vital aspect in knowledge creation in teaching. When we discover analogies, our knowledge expands rapidly and more assuredly.

---

[23] Hofstadter (1979), Chap. 2, Sect. Isomorphisms induce meaning.
[24] Shannon (1948).

The best studied axiomatic system is the theory of numbers, i.e., arithmetic, and Table 3.6 immediately tells us that any system that follows the template of the axiomatic (formal) system can be isomorphically translated into an arithmetical system and hence computerized and run on a computing system. It is therefore a matter of time that we will one day learn to restate biological systems in terms of arithmetical rules and computerize the whole system by analogy. Mathematicians would then be reading the book of life, developing not just vaccines and drugs but also species of their own design. This is not some pipedream of a mathematician playing God, but the untapped potential hidden in mathematics, which itself began with the instinctive ability to count with numbers 0, 1, 2, 3, more to 0, 1, 2, ... (without limit); to add and subtract (credit and debit); to quantify, measure, and compare; to put things and action steps in correspondence with numbers, give them numerical identity labels; etc., till counting and computing became the very essence of our existence and prosperity. It is now ready to overwhelm human intelligence and rob billions of humans of livelihood and hence of their life.

To a human a new interpretation is knowledge gained if it accurately reflects some isomorphism in the real world. When different aspects of the real world are isomorphic to each other one single formal system can be isomorphic to both and therefore hide two passive meanings. This kind of double-valuedness (or even multi-valuedness) of symbols and strings is an extremely important phenomenon. They provide us with deep understanding. Note also that reality and formal systems are independent. No one needs to be aware that there is an isomorphism between the two. Note that much of the mathematics used by physicists was not created for use by physicists! The mathematical scaffolding happened to be around when the physicists needed it to hang their physical observations on it. Mathematical structures are like mannequins on which we try out our conjectured laws of Nature.

Do formal systems throw light on truths in the domain of its interpretation? When one glimpses an isomorphism it is natural to wonder and explore the extent to which reality mimics the formal system. Conversely, can all of reality be turned into a formal system? *Nobody knows*. Mathematical modeling is all about such explorations.

Thus a basic belief in group theory led to the current models of particle physics. However, the fact remains that unless we already know that all the theorems of the formal system express truths under our chosen interpretation, we will not be sure [see Fig. 3.9 (right)]. That is, we must know all about the formal system as well as the interpreted system. Alas, this is impossible since in advancing knowledge, most of our intelligence is used in making conjectures and strenuously attempting to refute them forever to discover if our conjectures are flawed. To mere mortals, truth is not knowable.

Discovering unexpected isomorphisms is a genuine intellectual act. When an abstract symbol is paired with a word that stands for something in the real (or an imagined) world, we call the symbol-word correspondence *interpretation*. Once the correspondence is made, the abstract theorems can be interpreted to provide true statements in the chosen world. The process involves trial-and-error based on educated guesses. Then suddenly the Erdös-Renyi theorem kicks in, serendipity plays an active role, and everything appears to fall in place as a large blob emerges and the tip of a potential isomorphism becomes visible. Curiosity drives us on to discover the extent of the emerging isomorphism. This is tricky since abstract axiomatic systems deal with form and not with meaning. The symbol strings they generate do not carry any meaning. Humans add meaning to strings by creating isomorphisms. While we can create axiomatic systems and then look for real life isomorphisms, the reverse can also be done, i.e., we discover axiomatic systems taking cues from reality. Modern physicists do both. Much of their success depends on discovering isomorphisms.

Let me reemphasize: reality and formal systems are independent. No one needs to be aware that there is an isomorphism between the two. Much of the mathematics used by physicists was not created for use by physicists! Physicists found it lying around when they needed it. Mathematicians did not create mathematics so that physicists could use it centuries later! Small wonder then that physicists are so amazed at

The power and unreasonable effectiveness of mathematics in the natural sciences.[25] —
Eugene P. Wigner

Mathematics is abstract; it ignores meaning or context. It is a language we cleverly use to draw analogies between contexts by discovering shared concepts among them. Some examples are shown in Fig. 3.10 where we have only two symbols, a white dot and a colored dot arranged in a regular matrix of dots ordered in regularly spaced horizontal rows and vertical columns. See the dot matrices carefully. Are you seeing dots or giving it a meaning or even jumping to conclusions? What was the context or the world you inadvertently assumed should be applied to the dot matrices? What was the basis of your assumption? How would a blind person using Braille interpret the matrices? Does ignorance amount to blindness?

Those with an unimpaired vision, very likely saw, from left to right in the top row a fern, either a young lady or an old woman, either people going down or climbing up the stairs, an ant moving inside or outside a cage, and a musician or the face of a girl. Why is it that the same dot matrix (except for the fern) cause such irreconcilable interpretations? Why is it that given the same country, two political parties interpret the same facts so differently and so do the voters who vote for their candidates in elections? It is so because genuinely different interpretations are possible. Interpretation is in the mind, not in the representation. Even when we agree completely on facts, and solemnly abide by the agreed upon rules of the game, we can still vehemently disagree on interpretations as opposing parties do in court cases.

The second row in Fig. 3.10 tells us that there are additional pitfalls in the game. The way we represent (script) symbols (individually or in groups) itself may cause

---

[25] Wigner (1960).

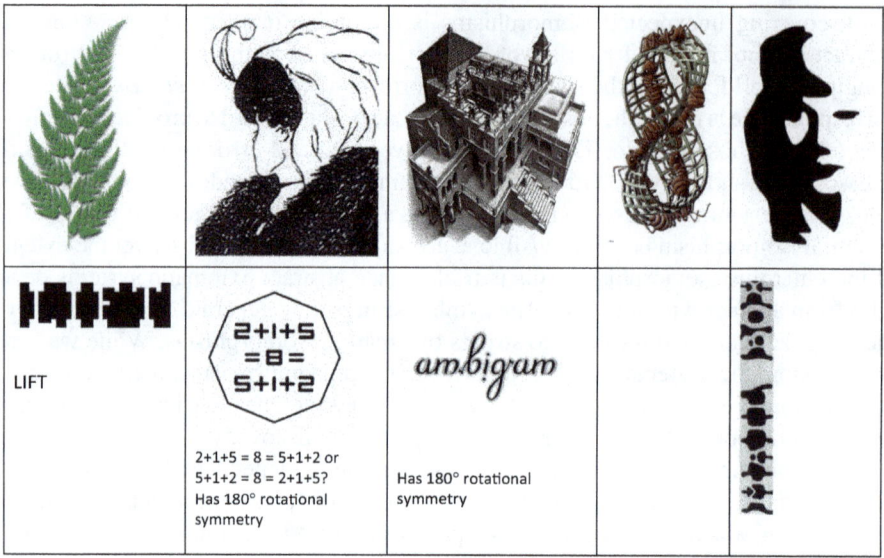

Fig. 3.10 Ambiguity. In real life we are often faced with the problem of choosing a point of view. *Source* (Top row, left to right) Fern, https://commons.wikimedia.org/wiki/File:Fraktal_ormbunke.jpg (Public domain). My wife and my mother-in-law, by the cartoonist W. E. Hill, 1915 (adapted from a picture going back at least to a 1888 German postcard). https://commons.wikimedia.org/wiki/File:Youngoldwoman.jpg (public domain). Ascending and descending, 1960. https://www.wikiart.org/en/m-c-escher/ascending-descending (Fair Use). Mobius Strip II—Red Ants (1963) by M. C. Escher. The Möbius Band in Art. https://sites.google.com/site/themobiusbandart/artistic-implications (Fair use). A musician or a girl's face? https://www.youtube.com/watch?v=eTFVlQsI9OU. Lower row: Ambigrams (from left to right). Lift. $2 + 1 + 5 = 8$ https://commons.wikimedia.org/wiki/File:Ambigram-8-eight-math-2-1-5-rotation-mirror-basile-morin.gif (in public domain); see also: https://www.slideshare.net/norainisaser/amazing-optical-illusion. Ambigram, https://commons.wikimedia.org/wiki/File:Ambigram_of_the_word_ambigram_-_rotation_animation.gif This file is licensed under the Creative Commons Attribution-Share Alike 4.0 International license. Mail Box (Figure and Ground) Hofstadter (Paperback, Vintage 1989), p. 67. (Fair use)

genuine contradictions, ambiguity, and unresolvable recognition issues, etc. The word ambigram was coined by Douglas Hofstadter (author of the classic book Gödel, Escher, Bach). An ambigram is a calligraphic design that has several interpretations as written.

> An ambigram is a visual pun of a special kind: a calligraphic design having two or more (clear) interpretations as written words. One can voluntarily jump back and forth between the rival readings usually by shifting one's physical point of view (moving the design in some way) but sometimes by simply altering one's perceptual bias towards a design (clicking an internal mental switch, so to speak). Sometimes the readings will say identical things, sometimes they will say different things.[26] — Douglas Hofstadter

---

[26] As quoted in Polster (n.d.) citing Douglas R. Hofstadter, Ambigrammi (in Italian), Hopefulmonster Editor, Firenze, 1987.

3.3 Universal Turing Machine

The whole of mathematics is structured around what we call axiomatic systems, just as nations lawfully organize themselves around a constitution. The structure of an axiomatic system (see Sect. 3.2) begins with a specification (a set of symbols and a grammar or typographical rules for combining the symbols into statements) regarding the construction of syntactically correct statements (well-formed formulas), a set of axioms (propositions regarded as self-evident truths or postulates; they may be any number including zero and infinity), and a finite set of inference rules that allow theorems to be generated using the axioms and previously derived theorems.

Creating an axiomatic system seems to require highly intelligent and creative humans. Deriving theorems within the system with specific non-trivial goals in mind (or developing algorithms) also seems to be a highly intelligent act. An algorithm, once developed, can always be mechanized.

Paradoxes

There is a fundamental difference between a straight line and a circle. The first has a beginning and an end, the latter has neither. Likewise, there are statements that are probably true or false in a given axiomatic system if we proceed in a linear manner, and there are statements for which the search for proofs tempt us to take circular paths in a spiraling way (say, with some parameter modification occurring in each round as in iterative methods). When circularity enters a proof process, one should be extraordinarily wary. Here is an example.

Consider the statement at the top in Fig. 2.1, Chap. 2, "I am a liar." It immediately strikes us as a grammatically correct statement in English. But can it be meaningfully right? How can anyone decide if I am telling the truth? If I am a liar, how can I be telling the truth? If I am not a liar, then why am I lying? If I say it under oath in a court of law, will it be valid evidence? This famous paradox in logic is called the Epimenides paradox after the Cretan philosopher Epimenides of Knossos (alive circa 600 BC) who is credited with the original version of the statement. Epimenides was a Cretan who made one immortal statement: "All Cretans are liars." Thus an axiomatic system may produce paradoxes via self-reference. It infects arithmetic because it is based on logic. This is why attaching meaning to arithmetical statements is so dicey because detecting paradoxes usually requires a high level of intelligence. That is why inductive reasoning must be used with great caution.

Statistical correlations are another dicey aspect of inductive reasoning. They do not necessarily establish cause-effect relationships. For example, if gin mixed with water makes you tipsy and so does whisky mixed with water, it does not mean water is the reason for your tipsiness. We need to be sure that a given correlation cannot be disentangled before it can be declared as a serious candidate as a law of Nature! Humans too often fail "to resist the urge to find causal relationships," even when they do not exist. This sometimes happens in clinical trials meant to evaluate the effectiveness and safety of medications or medical devices by monitoring their effects on large groups of people. The temptation to see causal relationships is high where the R&D costs for developing new drugs and vaccines are astronomical, success in marketing them is uncertain, and the clinical trial shows that the new treatment is neither clearly superior nor clearly inferior to standard treatment.[27]

There is another important aspect of knowledge exploration.

> [I]n making sense of the world, acting intelligently, and solving problems creatively, we do not rely solely on our mind's internal resources. Instead, we constantly have recourse to a vast array of culturally and socially embodied idea-spaces ... in forms as various as myths, business models, scientific paradigms, social conventions, practices, institutions, and even computer chips [to which] we have progressively offloaded ... for the sake of simplifying the burden on our own minds of rendering the world intelligible. Sometimes the space of ideas thinks for us. -- Richard Ogle.[28]

The discovery that all of mathematics can be codified in formal theories was the big event of the twentieth century. The advent of predicate calculus and the digital computer has profoundly influenced the way we view mathematics. Axiomatized mathematics is only a formal game, concerned solely with algorithmic manipulation of symbols. Mathematics is nothing but a kind of blind calculation.[29] The main benefits of axiomatization of mathematics are that axioms allow us to remove inaccuracies, ambiguities, and paradoxes that sometimes arise from excessive reliance on our intuitions. Axiomatization also permits a detailed analysis of relations between the fundamental principles of a theory (to establish their dependency or independence, etc.) and between the principles and theorems. Such investigations sometimes pave the way for further generalization of existing theories or integration of several specific theories. So far it has not been possible to eliminate intuition completely in constructing axiomatic systems, nor to affirmatively answer, in general, the question "Is a given proposition provable or not?" in a given axiomatic system. An axiomatic system creates an inventory of provable formulas or theorems. Axioms transcend reasoning, they depend on faith.

> Faith is an oasis in the heart which will never be reached by the caravan of thinking.—Khalil Gibran

Despite its deficiencies and because we are unable to do anything better, we rely upon axiomatic systems to make dependable predictions or understand the exploratory steps that led to our present state of knowledge or predicament! In the process we have learnt to dignify machines with intelligence.

---

[27] See, e.g., Rothwell (2005); Gludd (1999); Science (2012). "The largest comprehensive analysis of Clinicaltrials.gov finds that clinical trials are falling short of producing high-quality evidence needed to guide medical decision-making." See also: Tonkin (2000).

[28] Ogle (2007).

[29] Such a viewpoint was adopted by Niels Bohr with respect to quantum mechanics. The viewpoint eventually came to be known as the Copenhagen interpretation of quantum mechanics.

3.4 Futility of Rote Education

The time has come to realize that rote education is futile because deep down

> Mathematics is a language plus reasoning; it is like a language plus logic. Mathematics is a tool for reasoning. — Richard P. Feynman. The Character of Physical Law. BBC. 1965.

> [A] scientific theory is a computer program that calculates the observations, and that the smaller the program is, the better the theory. If there is no theory, that is to say, no program substantially smaller than the data itself, considering them both to be finite binary strings, then the observations are algorithmically random, theory-less, unstructured, incomprehensible and irreducible. — Gregory Chaitin[30]

Since any axiomatic system can be arithmetized it means two things. (1) The information content of any axiomatic system can be translated into arithmetical form without any distortion and vice versa; and (2) therefore any axiomatic system can be simulated on a Universal Turing Machine (UTM). This means knowledge creation and computing are two faces of the same coin. This makes AI possible. With AI as the driving force, we expect a sharp reallocation of economic activity dominated by high-markup firms that pioneer AI.

> While economic observers have long worried about the growing dominance of Big Tech, few have apprehended the sheer scale of the problem. Today's technologies have handed exorbitant market power to dominant players across all sectors, to the detriment of the vast majority of people.[31] — Jan Eeckhout

The reality as perceived by De Loecker, Eeckhout, and Unger is:

> Thriving competition between firms is a central tenet of a well-functioning economy. The pressure of competitors and new entrants leads firms to set prices that reflect costs, which is to the benefit of the customer. In the absence of competition, firms gain market power and command high prices. This has implications for welfare and resource allocation. In addition to lowering consumer well-being, market power decreases the demand for labor and dampens investment in capital, it distorts the distribution of economic rents, and it discourages business dynamics and innovation. This has ramifications for policy, from antitrust to monetary policy and income redistribution.[32]

Where we differ is that while demand for labor will decrease and the need for investment capital will too, the important difference overlooked by the authors is that much of the seed capital will be cerebral rather than monetary and the workforce will be a network of digital devices spread globally energized by solar energy. The distortions will be much larger and radical; AI will boost innovation, and 3D manufacturing will facilitate small scale manufacturing that is local (including households) and available on demand. Manufacturing will undergo major restructuring in terms of location, funding, marketing, and innovation.

---

[30] Chaitin (2003).
[31] Eeckhout (2021).
[32] De Loecker et al. (2020).

The immediate future will belong to those who can converse with AI machines intelligently in conceptual terms, not to those who entertain themselves with irresponsible, biased, selective, salacious, and gossipy media. The latter infect and spread information as rumors, the former is expected to act like vaccines to prevent rumor mongering.

3.4.1 Exponential Growth is Our Undoing

The green line in Fig. 3.11 (left) shows the behavior of an exponential (e^x) curve. It grows very slowly and linearly at first, speeds up as it nears the knee of the curve, and then rapidly shoots upward. With a rising population (see Fig. 3.3a) one should expect mass unemployment soon that cannot be alleviated by welfare schemes while AI takes away jobs from humans. The millennials were born as we were rapidly approaching the knee of the curve in STEM advances that powered global socio-economic growth along with the exponential growth of computing power [Moore's law, Fig. 3.11(right)]. We are now in the knee of the curve and heading for even faster AI growth. It means that machines are rapidly outpacing humans in intelligence. Here is the smoking gun evidence.

Arguably, when IBM's Deep Blue computer defeated chess champion Garry Kasparov in game one of a six-game match on 10 February 1996, a threshold in AI was crossed. While Kasparov won the 6-game match on this occasion, he lost to IBM's Deep Blue supercomputer on 12 May 1997 in a 6-game rematch. Since then AI machines have been beating human champions in games left, right, and center: they beat human champions in Jeopardy (February 2011), the Chinese game Go (March 2016), Poker (January 2017), once again in Go by AlphaGo Zero (October 2017; it learnt on its own from a blank slate), again in chess (December 2017; the

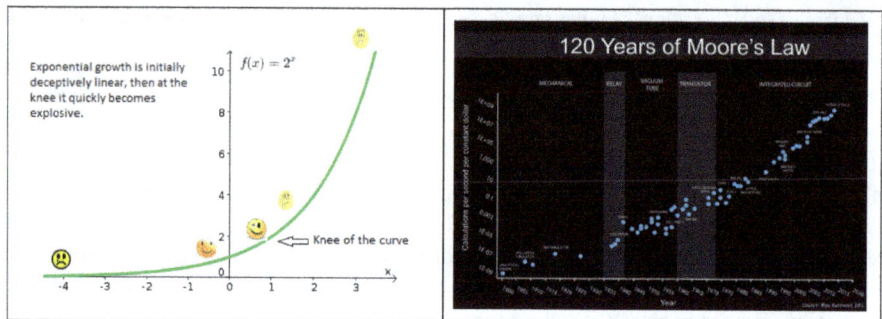

Fig. 3.11 Exponential. (Left) Exponential growth of the exponential function. *Source* Author. (Right) Exponential nature of Moore's Law plotted on semi-log scale (where it appears as a near linear curve). *Source of figure* Steve Jurvetson. An updated version of Moore's Law (based on Kurzweil's graph). Wikimedia Commons. https://commons.wikimedia.org/wiki/File:Moore%27s_Law_over_120_Years.png

machine taught itself in four hours), etc. And this is only the tip of the iceberg. Of these, the most significant is AlphaGo Zero which learnt purely by playing against itself millions of times over. It began by placing stones on the Go board at random but swiftly improved as it discovered winning strategies. It is a big step toward building general-purpose learning algorithms.

3.4.2 Knowledge Acquisition Requires an Environment

The future will belong to those who develop knowledge and learning skills, learn the art of protecting and licensing intellectual property generated from research, technology development, and encoding it in software, and attracting collaborative human talent. All this will require developing a new environment for a new era dominated by AI. AI brings in creative destruction as well as construction and hence sets up a new predator–prey game, the essence of which is captured in the logistic map.

The rapid and massive scale of COVID-19 spread has irreversibly affected our socio-economic environment in which knowledge is created, acquired, disseminated, and creatively used. Since COVID-19 surfaced in December 2019, a stark reality has emerged—the world is thoroughly underprepared to deal with it. The globally entrenched inequities—including the resources needed to collect timely and accurate data on death—are glaringly visible. We are not yet done with COVID-19. There is also the emerging toll on mental health, specially among teens. Ironically, the vaccines too lack a sound basis and are in use because desperate times require desperate measures. The long-term and even short-term adverse effects of these vaccines are unknown. Even more ironically, till a few years ago superpower summits were about managing nuclear weapons fueled by advances in quantum physics to keep the world safe, now it is about cyberweapons fueled by artificial intelligence that is eagerly seeking to harnesses the computing power of quantum computers. The world lives in trepidation. To spend or not on rote education is the question.

> The mind is not a vessel to be filled but a fire to be kindled.
> — *On Listening to Lectures*, Plutarch
> The limits of my language mean the limits of my world.
> — *Ludwig Wittgenstein*
> The woods are lovely, dark and deep,
> But I have promises to keep,
> And miles to go before I sleep,
> And miles to go before I sleep.[33]
> — *Robert Frost*

---

[33] Robert Frost. "Stopping by Woods on a Snowy Evening" from *The Poetry of Robert Frost*, edited by Edward Connery Lathem. Copyright 1923, © 1969 by Henry Holt and Company, Inc., renewed 1951, by Robert Frost. Reprinted with the permission of Henry Holt and Company, LLC.

References

Almécija S (2021) Fossil apes and human evolution. Science 372(6542):eabb4363. https://doi.org/10.1126/science.abb4363, https://science.sciencemag.org/content/372/6542/eabb4363?rss=1
Bera RK (2021) COVID-19 viewed from a different lens. SSRN. https://papers.ssrn.com/sol3/papers.cfm?abstract_id=3902583
Bernstein WJ (2004) The birth of plenty: how the prosperity of the modern world was created. McGraw-Hill
Chaitin GJ (2003) Leibniz, information, math and physics. arXiv:math.HO/0306303 v2. http://arxiv.org/abs/math/0306303
De Loecker J, Eeckhout J, Unger G (2020) The rise of market power and the macroeconomic implications. Quarterly J Econ 135(2):561–644. https://doi.org/10.1093/qje/qjz041
Eeckhout J (2021) Market power is eating the economy. Project Syndicate. https://www.project-syndicate.org/onpoint/high-stock-markets-reflect-market-power-no-competition-by-jan-eeckhout-2021-06
Gludd C (1999) Trials and errors in clinical research. The Lancet. https://www.thelancet.com/pdfs/journals/lancet/PIIS0140673699904592.pdf
Gödel K (1931) Über formal unentseheidbare Sätze der Principia Mathematica und verwandter Systeme I. Monatshefte für Mathematik und Physik 38:173–198 (Leipzig: 1931). (On formally undecidable propositions of Principia Mathematica and related systems I.) An English translation by B. Meltzer is available at http://jacqkrol.x10.mx/assets/articles/godel-1931.pdf. Part II of the paper was never published.
Grattan-Guinness (2000). Grattan-Guinness, I. A sideways look at Hilbert's twenty-three problems of 1900. Notices of the AMS, vol. 47, no. 7, August 2000, pp. 752–757. http://www.ams.org/notices/200007/fea-grattan.pdf
Hedges C (2003) What every person should know about war. The New York Times. https://www.nytimes.com/2003/07/06/books/chapters/what-every-person-should-know-about-war.html
Hofstadter DR (1979) Gödel, Escher, Bach: an eternal golden braid. Basic Books, New York
Holt (1964). Holt, J. C. How Children Fail. Penguin Education; 1964, p. 163. http://www.schoolofeducators.com/wp-content/uploads/2011/12/HOW-CHILDREN-FAIL-JOHN-HOLT.pdf
IPCC (2021) AR6 Climate change 2021: the physical science basis. IPCC. https://www.ipcc.ch/report/ar6/wg1/downloads/report/IPCC_AR6_WGI_Full_Report.pdf
Landauer R (1961) Irreversibility and heat generation in the computing process. IBM J Res Develop 5(3):183–191. Reprinted in IBM Journal of Research and Development 44(1–2):261–269. http://www.pitt.edu/~jdnorton/lectures/Rotman_Summer_School_2013/thermo_computing_docs/Landauer_1961.pdf
Laplace PS (1822) A treatise of celestial mechanics. translated from the French, and elucidated with explanatory notes by Henry H. Harte. Printed for Richard Milliken, Bookseller to the University; and for Longman, Hurst, Rees, Orme and Browne, London. https://archive.org/details/treatiseofcelest12lapl
LePan N (2020) History of pandemics. Visualizing the history of pandemics. Visual Capitalist. https://www.visualcapitalist.com/history-of-pandemics-deadliest/
Ogle R (2007) Smart world: breakthrough creativity and the new science of ideas. Harvard Business School Press. Quote reproduced from https://richardogle.typepad.com/site/smart_world_excepts/
Piketty T (2014) Capital in the twenty-first century. Harvard University Press
Polster B (n.d.) Mathemagical ambigrams. https://www.qedcat.com/articles/ambigram.pdf
Rothwell PM (2005) External validity of randomised controlled trials: "to whom do the results of this trial apply?" The Lancet. https://doi.org/10.1016/S0140-6736(04)17670-8
Science (2012) Large-scale analysis finds majority of clinical trials don't provide meaningful evidence. Science News. https://www.sciencedaily.com/releases/2012/05/120501162702.htm

References

Shannon CEA (1948) A mathematical theory of communication. Reprinted with corrections from the Bell System Technical Journal 27:379–423, 623–656. http://people.math.harvard.edu/~ctm/home/text/others/shannon/entropy/entropy.pdf

Tonkin AM (2000) Evaluation of large scale clinical trials and their application to usual practice. Heart, pp 679–684. https://www.researchgate.net/publication/12243548_Evaluation_of_large_scale_clinical_trials_and_their_application_to_usual_practice/link/0a85e53a4c25802d83000000/download

Turing AM (1936) On computable numbers, with an application to the Entscheidungsproblem. Proc London Math Soc S2-42(1):230–265, 1936–1937. https://www.cs.virginia.edu/~robins/Turing_Paper_1936.pdf Correction at: Turing AM On computable numbers, with an application to the entscheidungsproblem. A correction S2-43(1):544–546, 1938. http://www.turingarchive.org/viewer/?id=466&title=02

Wigner EP (1960) The unreasonable effectiveness of mathematics in the natural sciences. Richard Courant lecture in mathematical sciences delivered at New York University, May 11, 1959. Commun Pure Appl Math, 13:1–14. https://www.maths.ed.ac.uk/~v1ranick/papers/wigner.pdf

Chapter 4
Evolving AI Raises Human Creativity Concern

Abstract The exponential rise of science, technology, engineering, and mathematics (STEM) since the 1900s has completely changed the socio-economic context in which the Patent Act of 1790, and its successive amended versions were enacted. Since then the person of ordinary skill in the arts (PHOSITA) and in relation to this hypothetical person, the meaning of utility, novelty, non-obviousness, of an invention requiring human ingenuity and the manner in which the invention is to be disclosed to the public in exchange for a limited period monopoly over the invention by the inventor has undergone a sea change. In the last few decades, the world has seen a dramatic change in socio-economic-political structures, remarkable advances in STEM, e.g., in information and computing technologies, quantum computing, genetic engineering and synthetic biology, artificial intelligence, etc. These have had an enormous impact on the environment in which the *Homo sapiens* find themselves in. Such drastic changes are harbingers of natural speciation, an event that may not be too far off with unknown consequences. The species that succeed the *Homo sapiens* will likely be so far superior in intellect, intuition, and serendipity as to drive the *Homo sapiens* to extinction. This book assumes such an unfolding scenario and therefore suggests interim changes to the patent system so that the present debilitating stresses it faces, especially in the form of litigation, are substantially reduced. Our successor species will then perhaps remember us based not solely on our fossil record but also on our ability to anticipate the future and prepare for it intelligently.

Keywords Artificial intelligence · *Homo sapiens* · Intellectual property · Inventions · STEM

4.1 Introduction

Let us recall a few salient points from Chap. 2, Sect. 2.4.2. We, the *Homo sapiens*, have been around for about 300,000 years.[1] Records of our civilization date back approximately 6000 years. Some 12,000 years ago, we transitioned from a nomadic hunter-gatherer life to a pastoral-agricultural life. Society structured itself into families; women took care of the household; men earned a livelihood. Then from about 1500 AD to the later half of 20th century, an industrial economy developed with an accelerated growth of industrial activity and mechanization of agriculture. In a span of five centuries, we propelled the economy forward in graduated steps progressing from animal power to steam power to fossil fuel power to electrical power and now to digitized information power. With each new power source, society expansively restructured itself into increasingly complex and larger communities—extended families, cities, nations, alliances, institutions, modes of governance, dominions, etc.—and accordingly economies expanded their reach and scope from family businesses run locally to multinational corporations operating globally, employing millions of men and women. Women too began to enter the power corridors of corporations, and competing with men in all spheres of life—business, politics, arts, science, etc. Since the late 20th century the world began to transform at an unprecedented rate into a post-industrial economy which expects to ram its way to another future using wind, solar, and information power. This, in brief, is the progress of human civilization to date.

The post-industrial economy is rapidly enhancing its AI-driven knowledge-intensive services. The skill levels now required of even entry-level jobs for humans are higher than before and available jobs are less numerous. The middle class that bloated rapidly in the industrial age is now deflating equally rapidly as AI machines suck humans out of jobs. Ironically, while machines are neither looking for jobs nor do they need one, they have the power to snatch middle-class jobs because exceptional men configured machines to compute, while Nature has configured most other men for mere procreation. The industrial revolution and the patent system brought the pleasant prospect of improving humanity's collective lot for centuries to come. The AI revolution has shattered that prospect. Now only a few can live in style amidst machines, while the rest will sink into poverty for want of a job. Mr. Average is on his way to extinction.

Patent offices around the world are now well aware of this because they aggressively fuelled the patent system to its present state. Yet, no patent system in the world, till now, anticipated, much less understood the consequences of the exponentially accelerating advances in information and computing technologies and parallel advances in genetic engineering and synthetic biology. Their confluence within an amazingly short span in recent human history has caught them by surprise, even though Ray Kurzweil, a prominent inventor of our times, a recipient of the National Medal of Technology (1999), futurist and entrepreneur, has been alerting the world

---

[1] Hublin et al. (2017), and Richter et al. (2017). Prior to these papers, *Homo sapiens* were said to have been around for about 200,000 years.

4.1 Introduction

for over two decades about the rising power of AI.[2] Based on his well-founded "law of accelerating returns" Kurzweil has forecast that

> 2029 is the consistent date I have predicted for when an AI will pass a valid Turing test and therefore achieve human levels of intelligence. I have set the date 2045 for the 'Singularity' which is when we will multiply our effective intelligence a billionfold by merging with the intelligence we have created.[3]

I differ from Kurzweil in one important respect—his optimism that humans will benefit from such a development, rather I believe they will cross a threshold of speciation and will eventually become extinct. So, has the time come to sunset the patent system because our successor species will not only be dramatically different from us, but may eventually preside over our extinction to prevent us from competing for natural resources they need? Our legacy will not be our fossil record, but our amazing science, technology, engineering and mathematics (STEM) record for successor species to peruse and perhaps marvel.[4]

However, if the context does not change and we continue to live in a world where a sustainable population of *Homo sapiens* continue to dominate life on Earth, even then we will need to revisit all aspects of the patent system: (1) the legal necessity to adapt the system to align with the needs of today; (2) the need to overhaul the patent act; (3) the need for a novel appellate system to arbitrate on alleged patent invalidity, infringement, etc., and (4) above all a new definition of the inventor and his invention.

The patent system was intended to be a judicious interplay between law and economics. That interplay has become increasingly murky. The modern world cannot be understood, let alone governed, without a multidisciplinary perspective held by all actors in the game. STEM has advanced so rapidly in the past few decades and so expansively that it has forced many middle-class families to rearrange their priorities, finances, and lives. Success in life will now depend even more on meritocracy than on inherited wealth. And the required education for success must be expensively paid for.

Once again, there is growing inequality globally. It would be wrong to look back at history and believe the past can point a way out of this inequality simply because

> History does present one clear-cut case of an orderly recovery from concentrated inequality: In the 1930s, the U.S. answered the Great Depression by adopting the New Deal framework that would eventually build the mid-century middle class. Crucially, government redistribution was not the primary engine of this process. The broadly shared prosperity that this regime established came, mostly, from an economy and a labor market that promoted economic equality over hierarchy—by dramatically expanding access to education, as under the GI Bill, and then placing mid-skilled, middle-class workers at the center of production.[5]

---

[2] Kurzweil (1999) and Kurzweil (2005).
[3] Reedy (2017). See also: Kurzweil (2010) and ITU News (2019).
[4] Bera (2019).
[5] Markovits (2019).

The goal of modern AI is to augment human intelligence. Since the 1930s, it has strived to facilitate humans to do superintelligent work. Mechanization augmented human brawn thus providing humans more opportunities to exploit the power of the human brain by climbing over the brawn barrier. AI now augments the human brain but the *Homo sapiens* have no further opportunity to climb the brain barrier but to speciate to a new species with superior mental characteristics. That new species will likely have human-style innovation and serendipity built into its genes. The patent system and the *Homo sapiens* will then be on their way to a Darwinian extinction. Even without that, on an ominous front, new weapons like hypersonic missiles, artificial intelligence, and cyberattacks are already altering global power dynamics. The *Homo sapiens* may destroy themselves even before Nature has time to do so. The following observations tell us a lot about our in-built limitations.

> [O]ther transformational technologies, such as railroads, electricity, radio, television, automobiles and airplanes, all took several decades before they reached that comparable level of ubiquity. Society had the time to sort out the norms, rules and laws governing those technologies and the respective roles of government and the private sector.[6]

To add to our woes, *the Economist* notes,

> The "internet of things" has been gathering pace for years, but the revolution is about to go into overdrive. By 2035 the world could have a trillion connected computers, built into everything from clothes to cows. This will bring gains that are individually small but they will be compounded again and again across the economy. As the internet becomes all-pervasive, ever more companies will act like tech companies—as all-conquering "platform" monopolies, for instance, or adherents of the data-driven approach that critics call "surveillance capitalism". Arguments about ownership, data, privacy, competition and security will spill over from the virtual world into the real one.[7]

In its European edition, *The Economist* notes,

> The "single market", once breathtaking in its ambition to eliminate all internal EU barriers for goods, services, capital and people, has failed to keep up with the economies it was trying to shape. Europe's economy is losing ground to global rivals. A decade ago ten of the world's 40 largest listed firms by market value were based in the EU; now only two are—in 32nd and 36th place. Desperately few of the world's leading startups are European. If Europe wants to create prosperity and world-beating firms, it needs not just to reinvigorate the single market, but also to rediscover that original vision.[8]

AI augmented surveillance technology too is spreading rapidly, and not just to illiberal governments but also in liberal democracies and autocratic states. Presently Chinese and US companies are the largest global suppliers of this cutting-edge technology. In the wrong hands, it is a dystopian technology. The future is already here and we are struggling to keep up. The middle class is on the brink of an economic slide and that can only mean a disaster for humankind. The comity of economists didn't

---

[6] Gerstell (2019).

[7] See Economist (2019a). The quote is from a summary emailed by the Economist on 12 September 2019 to a list of recipients it maintains.

[8] See Economist (2019b). The quote is from a summary emailed by the Economist on 12 September 2019 to a list of recipients it maintains.

see this coming! The global economic climate continues to increase in volatility. Some highly visible reasons include trade tensions between China and the USA, long-standing historical tensions in the Middle East, the disruption caused by Brexit, the rise of hardline economic nationalism, and the gathering force of populism. No one knows where humanity is heading; all we know is that the direction appears ominous. The economists are totally clueless; they are dazed and will remain dazed.

Also, we should not forget the unabashed, provocative, and predatory nature of China's theft of intellectual property rights enshrined in the USPTO granted patents.[9] This example alone should suffice to alert all countries about the vulnerabilities of national patent systems.

4.2 How Early Inventions Advanced Human Civilization

The emergence of *Homo sapiens* based on recent fossil records has been revised to date back to about 300,000 years ago.[10] The data further suggests that we didn't evolve only in East Africa. The timing, location, and behavioral changes of *Homo sapiens* since their emergence to their modern biology, especially the development of their brain and mind poses many intriguing questions. The emergence of our species and the Middle Stone Age appear close in time. However, innovations and inventions of technology had begun much earlier.[11] Two important examples are the Acheulean Handaxe (~1,700,000 years ago), first made by the hominin family about 1.76 million years ago in the Lower Paleolithic (a.k.a. Early Stone Age) and used well into the beginning of the Middle Paleolithic (Middle Stone Age) period, about 300,000–200,000; and the control of fire (800,000–400,000 years ago), possibly an invention of our ancestor, *Homo erectus* during the Early Stone Age (or Lower Paleolithic). Controlled use of fire ranks among the first great innovations. Fire is a source of light and heat to cook plants and animals, to keep predator animals away, to clear forests for planting, to heat-treat stone for making stone tools, and to burn clay for ceramic objects. Further, fires serve as gathering places, as beacons, and as spaces for ceremonial activities. The *Homo sapiens* (the only extant species of genus "Homo") invented art (~100,000 years ago; e.g., cave paintings); textiles (~40,000 years ago; e.g., the deliberate processing of organic fibers into containers or cloth); shoes (~40,000 years ago); ceramic containers (~20,000 years ago); agriculture (~11,000 years ago; "the partnership between plants and humans"); wine (~9000 years ago); wheeled vehicles assisted by draft animals (~5500 years ago; it provided the means to rapidly move abundant goods across a landscape, widened trade opportunities in terms of geography, customer base, craft specialization, collaborative exchange of technologies with distant competitors, etc.); chocolate (~4000 years ago); etc.

---

[9] Li and Alon (2019). See also: Bera (2019a).
[10] The original papers are Hublin et al. (2017) and Richter et al. (2017). See also: Callaway (2017). The earlier estimate of the emergence of *Homo sapiens* was approximately 200,000 years.
[11] See, e.g., Hirst (2019).

During this period, inventions were few, but crucial for the survival of *Homo sapiens* in a world that was "red in tooth and claw" and allowed only the fittest to survive. Darwin's theory of evolution of life[12] posits that all life, including the *Homo sapiens*, is related and that it descended from a common ancestor. It presumes life developed from non-life and that complex creatures evolve from less complex ancestors naturally through the random adaptation process of "descent with modification". Genetic mutations that survive in a given environment are passed on to the next generation while the weaker are eliminated from breeding. When enough beneficial mutations accumulate, a phase transition is triggered and an entirely different organism (not just a variation of the original) comes into existence. The fossil record provides overwhelming evidence in support of Darwin's theory.[13] It now appears that *Homo sapiens* are rapidly heading toward speciation triggered by advances in STEM-mediated AI.

Biological evolution is not about individuals, it is about inherited means of growth and development that enhances the survivability of a given species evolving in a given habitat. The most important characteristic of human evolution since the agricultural era is the emergence of an adaptive brain-brawn feedback system that continuously monitors and tweaks the environment to improve conditions for human survival. Humans have evolved not purely by physical adaptation in an environment but by also mentally adapting to and changing the environment by putting their mind to work in coordination with the rest of their body. Farming communities gradually developed groups with specialized roles, e.g., soldier, ruler, and finally the institution of individual property rights. When individual farmers began owning and cultivating their plots of land, it created a competitive spirit that greatly boosted efficiency and productivity. That competitive spirit and individual ownership rights also gave rise to protecting intellectual property (products of the mind rather than the brawn) to further boost societal efficiency and productivity. In 1500 CE the world's population was 458 million; in December 2022 it was 8 billion and growing. During this period the marketplace underwent a radical change:

> In the modern marketplace, knowledge is the critical asset. It is as important a commodity as the access to natural resources or to a low-skilled labor market was in the past. Knowledge has given birth to vast new industries, particularly those based on computers, semiconductors, biotechnology and designed materials.[14]

Along with this, the technology landscape too underwent a radical change:

> Western industrial technology has transformed the world more than any leader, religion, revolution, or war. Nowadays only a handful of people in the most remote corners of the Earth survive with their lives unaltered by industrial products. The conquest of the non-Western world by Western industrial technology still proceeds unabated.[15]

---

[12] Darwin (1859).

[13] See, *e.g.*, Dawkins (2010).

[14] Bloch (1990), p. 9. Quote as reproduced at Warshofsky (1994).

[15] Headrick (1981), p. 4.

4.2 How Early Inventions Advanced Human Civilization

It was so radical in that period that it accomplished the safe landing and walking on the Moon by two US astronauts, Neil Armstrong and Buzz Aldrin, on 20 July 1969 and their subsequent safe return to Earth on 24 July 1969, with the event being broadcast live on TV to a global audience. In Adam Smith's days (1723–1790) such a feat was possible only in science fiction, fairy tales, and mythology.

Contemporaneously, the information technology revolution had just begun and was poised to make an exponential climb in advancing technology propelled by Moore's law. The millennials would embrace the technology and the products it produced as if born to it. In 1999, Ray Kurzweil would write a prophetic book titled[16] *The Age of Spiritual Machines: When Computers Exceed Human Intelligence*. A quick glance at Chap. 3, Sect. 3.4.1 will convince one of the remarkable pace at which learning algorithms advanced within a quarter century beginning 1996. It was a "giant leap" for mankind in AI and an ominous sign for *Homo sapiens* whose core survival resource is its vaunted intelligence. These developments are related to deep data and deep learning (See Sect. 4.4).

AI-created unemployment will be massive, disruptive, and destructive. It will begin with the spread of mental illness, ill-feelings against the world, jealousy against the prosperous, the craving to destroy, etc. It is already visible in populist nationalism, intolerance against immigrants, and distrust of globalism. The most vulnerable will be those born into the middle class, brought up in cocooned security and the promise of a comfortable future if they did well in their rote education. When human intelligence is blunted by AI, and rote education disconnects with employment, the future will appear dystopian.

The evolving socio-economic patterns are discernible only when viewed over multiple generations. Rising prosperity since the industrial revolution blinded us. With every rise comes a fall, just like the swinging rise and fall of a pendulum ball. The downfall began about a decade ago. The economists did not see it coming. Their lack of deep STEM knowledge ensures that they will never be able to predict the future and hence their opinions, whatever they be, will be totally irrelevant and unwanted. For example, they did not foresee the impact of automation and the rise of AI and the consequences that may follow. As Zabal and Luria perceptively note:

> In retrospect, *the four and a half decades from 1933 to 1978 were a historical aberration.* The longer-term trend toward more inequality in capitalist economies, which prevailed before this period, has resumed after it. That leads us to conclude that *there may well be no technocratic or tax policy fix for capitalism's tendency to generate ever more inequality.*[17] [Emphasis in the original]

Further,

> The great exception in U.S. economic history, the brief period beginning in 1933, was not only an anomalous period of decreasing inequality but one that also established America's unrivaled economic and political dominance. That success was built upon a hard-to-reproduce, five-part perfect storm ...[18]

---

[16] Kurzweil (1999).

[17] Zabala and Luria (2019).

[18] Zabala and Luria (2019).

That storm included expropriation, demand stimulus, unionization, war production, and postwar economic hegemony. "No single economic, financial, or political factor in isolation explains the U.S. postwar economic trajectory ... that long-term trends toward inequality were only interrupted and then slowed by a surge in workers' bargaining power and a paring back of the wealth share of the top 1 percent of households."[19] The resurgence of income inequality is seen in the erosion of employee bargaining power vis-à-vis employers, and the inability of most newly minted college graduates to get and keep middle-class jobs. A downward socio-economic slide has begun. Much of the rise and concentration of economic power by the few drew momentum from the USPTO's lethargic recognition of the rapidly rising skill levels of the person having ordinary skill in the arts (PHOSITA) when granting patents. This flooded the market with unwarranted patents. Lacking STEM knowledge, the judiciary was helpless in bucking the trend and was reduced to entertaining and encouraging arguments over trivial matters as Burk and Lemley (2009) noted more than a decade ago:

> Despite repeated efforts to set out rules for construing patent claims ... parties and courts seem unable to agree on what particular patent claims mean. ... Literally every case involves a fight over the meaning of multiple terms, and not just the complex technical ones. Recent Federal Circuit cases have had to decide plausible disagreements over the meanings of the words "a," "or," "to," "including," and "through," to name but a few. ... Even after claim construction,[20] the meaning of the claim remains uncertain, not only because of the very real prospect of reversal on appeal but also because lawyers immediately begin fighting about the meaning of the words used to construe the words of the claims. ... Patent attorneys seize on such indeterminacy to excuse infringement or to expand their client's exclusive rights. ... [T]he patent system increasingly revolves around the definition of terminology rather than the substance of what the patentee invented and how significant that invention really is. ... [C]ourts define the scope of legal rights not by reference to the invention but by reference to semantic debates over the meaning of words chosen by lawyers.[21] (Footnote added.)

This degeneration has been gradual and irreversible. In practice, the patent document is seldom written for full comprehension by relevant technical experts (a judicially ignored violation of the patent act that requires full and comprehensible disclosure of the invention) but for lawyers who, in patent litigation, must present their client's case to generalist judges ignorant of the technical arts that support the patent.

Lee rightly notes:

> In an ideal world, personal biases would be irrelevant to judging. The job of a federal judge is to fairly apply the Constitution and federal statutes to particular cases. If the law were perfectly clear and unambiguous [axiomatized], it would be irrelevant who was put in charge of interpreting it. Of course, law doesn't actually work that way. Congress has defined patent law using general terms like "obvious," "novel," and "process." The courts give concrete meaning to those terms through a series of precedents. Hence, the biases of a court can have a significant impact on how the law is interpreted.[22]

---

[19] Zabala and Luria (2019).

[20] The process of determining how best to interpret the words that describe an invention in a patent in plain English.

[21] Burk and Lemley (2009).

[22] Lee (2012).

Thus, every 5–4 SCOTUS ruling is a lottery draw. And therein lies danger. Shapiro notes:

> I believe there is manifest danger in binding rulings, particularly in the field of patent law, made by courts that do not understand the issues before them. Justice Scalia's proclamation in Myriad[23] that the issues discussed were beyond the understanding of the court should raise serious red flags. Indeed, it is hard to imagine that any court, or system of law, can maintain institutional legitimacy, if it issues decisions that demonstrate misunderstanding of the field, or are not logically supported.[24] (Internal citation omitted.)

It is therefore imperative that issues related to patent validity and scope be decided by a novel Patent Validation Board (PVB) (see Chap. 5, Sect. 5.5.1) and not the courts.

In a consensus study, the National Academies (of the US) stated its position on the matter of "Facilitating Transdisciplinary Integration of Life Sciences, Physical Sciences, Engineering, and Beyond". Its abstract said,

> Convergence of the life sciences with fields including physical, chemical, mathematical, computational, engineering, and social sciences is a key strategy to tackle complex challenges and achieve new and innovative solutions. However, institutions face a lack of guidance on how to establish effective programs, what challenges they are likely to encounter, and what strategies other organizations have used to address the issues that arise. This advice is needed to harness the excitement generated by the concept of convergence and channel it into the policies, structures, and networks that will enable it to realize its goals.[25]

4.3 When STEM Changed the Face of Man-Made Inventions

Along with Ray Kurzweil's prophetic predictions about the future capabilities of AI machines—"By 2029, computers will have human-level intelligence"),[26] we recall from Chap. 2, Sect. 2.1.3, a salient observation from Richard Ogle in his book, *Smart World*: "Sometimes the space of ideas thinks for us."[27]

This smart world also faces unprecedented demographic changes due to variations in mortality, life expectancy, and a youthful population in countries where fertility is high. As Fig. 2.5 in Chap. 2 shows, the world is already heavily overpopulated. In the next three to four decades, the population of the more developed countries is likely

---

[23] SCOTUS (2013). Justice Scalia wrote: "I join the judgment of the Court, and all of its opinion except Part I–A and some portions of the rest of the opinion going into fine details of molecular biology. I am unable to affirm those details on my own knowledge or even my own belief. It suffices for me to affirm, having studied the opinions below and the expert briefs presented here, that the portion of DNA isolated from its natural state sought to be patented is identical to that portion of the DNA in its natural state; and that complementary DNA (cDNA) is a synthetic creation not normally present in nature".
[24] Shapiro (2015).
[25] NAP (2014).
[26] Fox News (2017).
[27] Ogle (2007).

to stagnate at about 1.3 billion. Their population is aging and would decline but for migration. The populations of Germany, Italy, Japan, and several states of the former Soviet Union that broke away are also expected to decline by 2050.[28] The world's flexibility to cope with such unprecedented socio-economic changes is untested.

Opportunities for digitizing and automating tasks are far from being over. The more cognitive tasks are automated and embellished with language processing and pattern matching, and enhanced with mechanized physical dexterity, mobility, and sensory perception, the bigger will be its impact on depriving human workers of jobs performing these tasks and on division of labor in a society. Eventually, *inter alia*, assembly line workers, taxi drivers and long-haul truckers (Google/Waymo, Tesla, nuTonomy, Uber, and many others have already invested in self-driving vehicles) will have to seek other forms of employment. Machines capable of perceptual tasks, e.g., language translation, speech recognition, text reading, computer vision will begin to replace human specialists in pathology, radiology, security, language translation, paralegal work, and many others.

Hardware for implementing AI software continues to progress rapidly as are energy sources that power the hardware. Beyond mere speed up and energy efficiency, other important technologies are advancing too. These include mobile Internet, IoT (Internet of Things), cloud computing and storage, AI, autonomous vehicles, robotics, virtual and augmented reality, virtual personal assistants, fitness trackers, everything cloud, 3D printing, hyperloop, drone services, renewable energy, and machine learning. And finally, there is bionics leading to a future population of superhumanoids potentially capable of forming a society and running the world.

Technology is marching ahead at an exponential rate. The perceived goal of AI researchers is no longer the mere mimicking of Mr average in intelligence but exceeding those of the best intellectuals. The millennials can only imagine how wildly disruptive their life will be as they grow older. Their generation will be disrupted by the fact that superintelligent machines can routinely conceive of ideas that no human being has entertained in the past, and that machines can invent sophisticated and advanced technological tools that will surpass anything *Homo sapiens* can create.

The millennials now face unprecedented survival challenges in the future. The most challenging of them will be their ability to adapt to the new world by competing against AI-embedded humanoids they are themselves creating. A biological evolution of intelligent life is waiting to happen, triggered by the *Homo sapiens'* quest to understand the universe not according to the scriptures but according to science. The *Homo sapiens* are on their way to becoming an endangered species within a century.

Imminent speciation may sound alarming, but it is scientifically plausible. The domesticated dog is a prime example that man is the instigating factor on Earth in changing the environment. It is he who could domestic a wild species through genetic breeding in a very small fraction of the time that Nature would have required. And now that man has learnt the secret of creating new species in a lab, the time is not too far when he would be doing it on a mass scale. The domestication of the gray wolf into dogs happened long before the Industrial Revolution (1760–1840), before literature

---

[28] See, e.g., UNPF (2017).

and mathematics, and before bronze, iron, and agriculture. This ancient partnership between man and animal entwined the fate of the two species. "The wolves changed in body and temperament. Their skulls, teeth, and paws shrank. Their ears flopped. They gained a docile disposition, becoming both less frightening and less fearful. They learned to read the complex expressions that ripple across human faces. They turned into dogs."[29] What will be our fate when the *Homo sapiens* speciate? What kind of territorial rights will we share with the new species? Who will be the master and who the slave?

Technology development began to rise sharply coinciding with the birth of the millennials. A glance at Fig. 2.5 in Chap. 2 (Population of the Earth), Fig. 3.3a in Chap. 3 (World population growth 1700–2100), and Fig. 3.11 in Chap. 3 (Exponential growth of exponential function) will provide an intuitive feel for the situation as it is developing for the millennials and their progenies. What needs to be noted is the growth of population and in parallel the growth of computer technology which is the key to AI and its future. AI is the substrate on which other technologies will grow.

A distinctive aspect of emerging technologies is their ability to create necessities or a 'must possess feel' not felt before. It is highly visible in myriad digital communication-plus devices ubiquitously available and affordable. Instant communication links that connect humans and devices via the Internet of Things (IoT) is now increasingly taken for granted. It has set in motion a disruptive restructuring of society by an "unseen hand" into a malleable global structure where people are tagged with a profile matrix that includes lifestyle, nationality, education, employability, religion, etc., usually in that order of importance. Society increasingly celebrates the individual than the family. The first millennials were born in the incipient stages of this disruption caused by the rapid automation of many hitherto human activities (physical and mental). It began affecting family structure and lifestyle, employment, skilling, and reskilling opportunities and the welfare management of a growing population of retirees who are culturally alienated from their millennial progenies. Compounding the problem, the progenies in turn face an uncertain and unpredictable job market. As AI advances, it diminishes employment opportunities, job stability and job quality for those without an elite education or highly specialized skills yet to be automated. Key factors in improving societal well-being are prosperity through gainful employment, individual well-being through good mental and physical health, sustainability of lifestyle through economic means and environmental upgradation, and justice and trust through ethical, moral and law enforcement standards. Recent events have shown that a societal phase transition is underway. The old socio-economic structure is crumbling, and a new stable structure is yet to take shape.

Phase transitions are disruptive. In socio-economics reliable mathematical models simply do not exist. Hence social "scientists" (economists in particular) are completely clueless in anticipating the disruptive nature of AI and the cataclysmic effect it will have on *Homo sapiens*. To understand the problem, one needs to understand what we mean by intelligence and knowledge and how the best of human minds use their brains (natural neural network) to create artificial neural networks

---

[29] Yong (2016).

to augment and amplify the power of the human brain. The earlier chapters of this book were meant to throw some light on these matters.

Man-made technology did a superb job in enhancing and embellishing human brawn power (e.g., in robotics, farming, transportation, civil construction, manufacturing, etc.); it is now trying to repeat it on human brain power. The first nugget of success came when Alan Turing proved that computing can be mechanized in toto![30] And he proved it using mathematics! Actual computer hardware came later notably with the ENIAC (Electronic Numerical Integrator And Computer)[31] in 1945, developed in the United States by Army Ordnance to compute World War II ballistic firing tables. Since then mechanized computing power has grown at the phenomenal pace captured in Gordon Moore's law[32] (see the log-linear relationship of transistor counts for micro-processors in Fig. 2.7 of Chap. 2). Ever since, doing arithmetic has ceased to be an intelligent activity (*creating* new mathematics still remains a highly intelligent activity). AI is now well on its way to mechanize an immense amount of "intelligent" activity that a few years ago created well-paid jobs on a large scale. AI is now poised to make millions of even well-educated people unemployable. The hardest hit will be those whose skill levels are so low that they can be replaced by machines at a low cost.

Using technology to improve well-being and smoothen disruptive change was a very fortunate outcome of the industrial revolution because the potential for improving skill levels and the means to do it was possible and affordable on a large scale to improve employee productivity. That is no longer the case in an AI-dominated environment, and this environment is rapidly expanding into the future where data and AI, and connectivity and platforms will be ubiquitous. Employability already demands innovation skills and labor fluidity to match demand with availability. With the global population bursting at the seams and performing at abysmal skill levels when pitted against advancing AI, only a miniscule fraction of the world's population can be gainfully employed. The rest will form a massive gig economy where the weak suffer what they must.

4.4 Beyond the Data-Driven World

Current AI research is centered around a set of computing technologies that tries to mimic how humans use their nervous system and sense organs to sense, learn, reason, and act. For example, deep learning is a form of machine learning based on layered representation of variables called neural networks. These are widely used in applications that rely on pattern recognition. Natural language processing (NLP) and knowledge representation and reasoning was used by IBM Watson to win the

---

[30] Turing (1936).
[31] Moye (1996).
[32] Moore (1965).

Jeopardy competition in 2011.[33] Watson used a series of complex search algorithms, and some heavy-duty computing power to find answers with the highest probability of being correct.[34] (I have reservations about pursuing NLP. What needs to be done is developing a language that does not admit ambiguity because each of its messages can then be pinned to a specific context through a web of connections.) With present-day NLP methods, when the context is narrow and unambiguous, it is powerful enough with potential for further growth in web searches, self-driving cars in low traffic, healthcare diagnostics and targeted treatments, etc. AI and robotics are now in wide use in agriculture, food processing, fulfillment centers, and factories where they accentuate jobless economic growth. Once a new, context discriminating, man–machine *lingua franca* is created for universal use, AI will advance exponentially. Beyond this lies the vastly challenging task of discovering axiomatic systems that can encapsulate massive amounts of as yet unconnected data/observations. Here is a profound insight from Gregory Chaitin we repeat from Chap. 3:

> [A] scientific theory is a computer program that calculates the observations, and that the smaller the program is, the better the theory. If there is no theory, that is to say, no program substantially smaller than the data itself, considering them both to be finite binary strings, then the observations are algorithmically random, theory-less, unstructured, incomprehensible and irreducible.[35]

This, I believe, describes the heart and soul of AI, i.e., a successful AI-system should be able to discover an axiomatic-system that looks at vast amounts of data, categorizes the data by doing a correlation analysis, creates a random set of samples and gleans common "features" among members of this random set, proposes a parsimonious set of axioms and rules of inference that would reproduce the observed features *and* predict new features as "theorems" where one is forever trying to see if $x = y$ or not (x and y are two validly constructed statements or axioms in the axiomatic system). Parsimony is the key attribute by which an AI system must be measured for its excellence and effectiveness. While seeking parsimony one should always bear in mind the impossibility of proving the consistency and completeness of any axiomatic system to which Gödel's theorems[36] apply, the limitations of the computing powers of a Universal Turing Machine,[37] the limitations imposed by the postulates of quantum mechanics[38] and that information is physical[39] and hence governed by the laws of physics.

I see AI at a juncture where physics was in the early 1900s when quantum mechanics burst on the scene. It put physics and our perception of Nature on an entirely different conceptual footing. Likewise, AI researchers must decide what

---

[33] Best (2013).
[34] Lynley (2011).
[35] Chaitin (2003).
[36] Gödel (1931).
[37] Turing (1936).
[38] Nielsen and Chuang (2000).
[39] Landauer (1961).

we mean by intelligence and cognition and thus what it means to be human in an AI-driven world. Our survival will depend on how we integrate with AI machines.

4.5 Speciation of the *Homo sapiens*

A breakthrough in technology is assured when synthetic biology, AI, and quantum computing eventually integrate into creating new life forms through forced speciation that will likely lead to a superspecies, the humanoid. This will likely make the *Homo sapiens* extinct (all earlier species, *e.g.*, *H. habilis*, *H. erectus*, and *H. heidelbergensis* as well as the Neanderthals (*H. neanderthalensis*), the early form of *Homo sapiens* called Cro-Magnon, and the enigmatic *H. naledi* in the *Homo* genus are now believed to be extinct[40]) or domesticated by them (as the wolves were domesticated into dogs). This will completely overturn all predictions about the rate and directions of AI advances that are currently favored.[41] As I have noted recently

> We envisage a world where genetic engineering, artificial intelligence (AI), and quantum computing (QC) will coalesce to bring about a forced speciation of the *Homo sapiens*. A forced speciation will drastically reduce the emergence time for a new species to a few years compared to Nature's hundreds of millennia. … Accelerating speciation mediated by *Homo sapiens* via domestication, gene splicing, and gene drive mechanisms is now scientifically well understood. Synthetic biology can advance speciation far more rapidly using a combination of clustered regularly interspaced short palindromic repeats (CRISPR) technology, advanced computing technologies, and knowledge creation using AI. The day is perhaps not far off when *Homo sapiens* itself will initiate its own speciation once it advances synthetic biology to a level where it can safely modify the brain to temper emotion and enhance rational thinking as a means of competing against AI-embedded machines guided by quantum algorithms.[42]

The repercussions of forced speciation will be enormous. The role of natural humans and humanity's faith in spirituality if humanoids take charge will undergo a sea change. Such a biological evolution of intelligent life, triggered by the *Homo sapiens'* curiosity-driven quest to understand the universe within a rational, axiomatized framework will force humanity to reassess the meaning of life, its place and significance in the universe, and above all its ability to merely survive, much less survive with dignity.

References

Bera RK (2019) Synthetic biology, artificial intelligence, and quantum computing. book chapter in genetic engineering technology and synthetic biology. In: Nagpal ML (ed) IntechOpen,

---

[40] Encyclopaedia Britannica (2019).

[41] See, e.g., Stone (2016).

[42] Bera (2019).

London, (to appear in 2019). Prepublication access to the chapter has been provided by InTech at https://www.intechopen.com/online-first/synthetic-biology-artificial-intelligence-and-quantum-computing; PDF at https://www.intechopen.com/chapter/pdf-download/65149

Bera RK (2019a) China: the trump factor revisited. Available at SSRN: https://papers.ssrn.com/sol3/cf_dev/AbsByAuth.cfm?per_id=1752696

Best J (2013) IBM Watson: the inside story of how the Jeopardy-winning supercomputer was born, and what It wants to do next. TechRepublic. http://www.techrepublic.com/article/ibm-watson-the-inside-story-of-how-the-jeopardy-winning-supercomputer-was-born-and-what-it-wants-to-do-next/

Bloch E (1990) Can the U.S. compete? World Link

Encyclopaedia Britannica (2019) The editors of encyclopaedia Britannica. Homo: hominin genus. Encyclopaedia Britannica. Last updated 09 Jan 2019. https://www.britannica.com/topic/Homo

Burk DL, Lemley MA (2009) Fence posts or sign posts? Rethinking patent claim construction. University of Pennsylvania Law Review, vol 157, No. 6. Symposium: the foundations of intellectual property reform, pp 1743–1799, http://papers.ssrn.com/sol3/papers.cfm?abstract_id=1358460

Callaway E (2017) Oldest *Homo sapiens* fossil claim rewrites our species' history. Nature. (Corrected 08 June 2017). https://www.nature.com/news/oldest-homo-sapiens-fossil-claim-rewrites-our-species-history-1.22114

Chaitin GJ (2003) Leibniz, Information, math and physics. arXiv:math.HO/0306303 v2. http://arxiv.org/abs/math/0306303

Nielsen MA, Chuang IL (2000) Quantum computation and quantum information. Cambridge University Press. [Errata at http://www.squint.org/qci/]

Darwin C (1859) On the origin of species by means of natural selection. First Edition. John Murray, London. http://darwin-online.org.uk/content/frameset?itemID=F373&viewtype=text&pageseq=1

Dawkins R (2010) The greatest show on earth: the evidence for evolution. Free Press, Reprint edition 2010

Economist (2019a) Now the world will change as computers spread into everyday objects. The Economist. https://www.economist.com/leaders/2019/09/12/now-the-world-will-change-as-computers-spread-into-everyday-objects

Economist (2019b) A singular opportunity. The Economist. https://www.economist.com/leaders/2019/09/12/a-singular-opportunity

Fox News (2017) Ray Kurzweil predicts computers will be as smart as humans in 12 Years. Fox News. http://www.foxnews.com/tech/2017/03/16/ray-kurzweil-predicts-computers-willbe-as-smart-as-humans-in-12-years.html

Gerstell GS (2019) I work for N.S.A. We cannot afford to lose the digital revolution. The New York Times. https://www.nytimes.com/2019/09/10/opinion/nsa-privacy.html

Gödel K (1931) Über formal unentseheidbare Sätze der Principia Mathematica und verwandter Systeme I. Monatshefte für Mathematik und Physik 38:173–198. (On formally undecidable propositions of Principia Mathematica and related systems I.) (Visit http://jacqkrol.x10.mx/assets/articles/godel-1931.pdf for an English translation by B. Meltzer.)

Headrick DR (1981) The tools of empire: technology and European imperialism in the nineteenth century. Oxford University Press, New York

Hirst KK (2019) Top 10 inventions in ancient human history. ThoughtCo. https://www.thoughtco.com/top-inventions-in-ancient-human-history-172900

Hublin J et al (2017) New fossils from Jebel Irhoud, Morocco and the pan-African origin of *Homo sapiens*. Nature 546:289–292. https://www.nature.com/articles/nature22336

ITU News (2019) The future is better than you think: predictions on AI and development from Ray Kurzweil. ITU News. https://news.itu.int/the-future-is-better-than-you-think-predictions-on-ai-and-development-from-ray-kurzweil/

Kurzweil R (1999) The age of spiritual machines: when computers exceed human intelligence. Viking Press

Kurzweil R (2005) The singularity is near: when humans transcend biology. Viking
Kurzweil R (2010) How my predictions are faring. Kurzweilai.net. http://www.kurzweilai.net/images/How-My-Predictions-Are-Faring.pdf
Landauer R (1961) Irreversibility and heat generation in the computing process. IBM J Res Develop 5(3):183–191. http://www.pitt.edu/~jdnorton/lectures/Rotman_Summer_School_2013/thermo_computing_docs/Landauer_1961.pdf
Landauer R (1991) Information is physical. Phys Today, 23–29
Lee TB (2012) How a rogue appeals court wrecked the patent system. Arstechnica. http://arstechnica.com/tech-policy/2012/09/how-a-rogue-appeals-court-wrecked-the-patent-system/2/
Li S, Alon I (2019) China's intellectual property rights provocation: a political economy view. J Int Bus Policy. (online). https://link.springer.com/article/10.1057%2Fs42214-019-00032-x, and https://doi.org/10.1057/s42214-019-00032-x
Lynley M (2011) Watson supercomputer defeated in Jeopardy by lone physicist—long live humanity! VentureBeat.com. https://venturebeat.com/2011/03/07/humanity-wins-against-watson/
Markovits D (2019) How life became an endless, terrible competition. The Atlantic. https://www.theatlantic.com/magazine/archive/2019/09/meritocracys-miserable-winners/594760/
Moore GM (1965) Cramming more components on to integrated circuits. Electronics 38(8). https://drive.google.com/file/d/0By83v5TWkGjvQkpBcXJKT1I1TTA/view
Moye WT (1996) ENIAC: the army-sponsored revolution. https://web.archive.org/web/20170521072638/http://ftp.arl.mil/~mike/comphist/96summary/index.html
NAP (2014) National research council. convergence: facilitating transdisciplinary integration of life sciences, physical sciences, engineering, and beyond. Division on Earth and life studies; board on life sciences; committee on key challenge areas for convergence and health. The National Academies Press, Washington D.C. https://www.nap.edu/catalog/18722/convergence-facilitating-transdisciplinary-integration-of-life-sciences-physical-sciences-engineering
Ogle R (2007) Smart world: breakthrough creativity and the new science of ideas. Harvard Business School Press
Reedy C (2017) Kurzweil claims that the singularity will happen by 2045. Futurism. https://futurism.com/kurzweil-claims-thatthe-singularity-will-happen-by-2045/
Richter D et al (2017) The age of the hominin fossils from Jebel Irhoud, Morocco, and the origins of the Middle Stone Age. Nature 546:293–296
SCOTUS (2013) Ass'n for Molecular Pathology v. Myriad Genetics, Inc., 133 S. Ct. 2107, 2111. http://www.supremecourt.gov/opinions/12pdf/12-398_1b7d.pdf
Shapiro Z (2015) Patent law, expertise, and the court of appeals for the federal circuit. Harvard Law. http://blogs.law.harvard.edu/billofhealth/2015/07/14/patent-law-expertise-and-the-court-of-appeals-for-the-federal-circuit/
Stone P et al (2016) Artificial intelligence and life in 2030. One Hundred Year Study on Artificial Intelligence: Report of the 2015–2016 Study Panel (2016). https://ai100.stanford.edu/sites/default/files/ai_100_report_0831fnl.pdf
Turing AM (1936) On computable numbers, with an application to the Entscheidungsproblem. Proceedings of the London Mathematical Society, S2–42(1):230–265, 1936–1937. https://www.cs.virginia.edu/~robins/Turing_Paper_1936.pdf Correction at: Turing AM (1938) On computable numbers, with an application to the Entscheidungsproblem. A Correction. S2–43(1):544–546. http://www.turingarchive.org/viewer/?id=466&title=02
UNPF (2017) World population dashboard. United Nations Population Fund. http://www.unfpa.org/pds/trends.htm
Warshofsky F (1994) The Patent Wars. John Wiley
Yong E (2016) A new origin story for dogs. The Atlantic Daily. https://www.theatlantic.com/science/archive/2016/06/the-origin-of-dogs/484976/
Zabala C, Luria D (2019) New gilded age or old normal? American Affairs. https://americanaffairsjournal.org/2019/08/new-gilded-age-or-old-normal/

Chapter 5
Vulnerabilities of the Patent System

Abstract Under prevalent patent systems around the world, patenting of inventions related to advances in quantum computing, synthetic biology, and artificial intelligence (AI) have begun to raise serious concerns. Advances in AI are particularly problematic because their influence will be felt on all hitherto patent eligible inventions. Since AI machines have the potential to prolifically invent patentable technology, it will undoubtedly shake the very foundation on which the patent system presently rests. It will require us to redefine what we mean by novelty, non-obviousness, and written description of the invention (e.g., shouldn't a binary string suffice as written description because it is the lingua franca of computers). In this chapter we focus on the US patent system.

Keywords Patent · Artificial intelligence · Jurisprudence · Rationalism

5.1 Stressed out Patent System

The entire US patent system is stressed out and functions with well-intentioned Band-Aids. The patent system needs to be redefined starting from what is a patentable invention, who qualifies as an inventor, what are the attributes of a PHOSITA (person having ordinary skill in the art) with respect to a given patent under examination and correspondingly what should be the qualifications of a patent examiner. It is now clear that the judiciary should not be involved in any aspect of patent litigation until the validity of the patent-in-suit is firmly established by an independent statutory body (we suggest a Patent Validation Board (PVB), see Sect. 5.5.1) whose decision will be final and binding on all. The patent law needs an overhaul given that a steady stream of inventions will now automatically flow from organizations even without incentives from its AI-savvy employees and AI-embedded machines. In the future, our primary source of inventions will be AI machines. Unlike the eighteenth century, the artisan inventor is no longer the prized source of inventions around whom the patent system was built. The modern inventor is a STEM researcher, savvy in using information technology, is part of a funded research team with the goal of building a patent portfolio for a corporation that needs to stay competitive in the marketplace and to weaponize itself against IP litigation. Several past judicial decisions, e.g., those related to patentable subject matter and the doctrine of equivalents must be revisited, and all such matters should be decided by the PVB. We also note that the judiciary itself will accordingly undergo a radical change in its functioning when AI begins to invade its portals.

Much of the confusion and consequent opportunistic patent litigation that arises today is due to the three judicially created exceptions to the U.S. Patent Act's broad patent-eligibility principles: 'laws of nature, natural phenomena, and abstract ideas' whose scope and limitations remain unclear and confusingly dealt with in litigation. From our modern understanding of physics, mathematics, algorithms, computations, life sciences, and information we conclude that a rigid adherence to the exceptions to maintain *stare decisis* in jurisprudence is irrational. It is also anachronistic because the judiciary lacks the deep understanding of STEM, which post-1900 has undergone dramatic changes. Consequently, the patent system is wading in a quagmire of its own making. In particular, the judiciary errs in believing that the laws of nature are known to mankind and therefore they are "part of the storehouse of knowledge of all men" and "free to all men and reserved exclusively to none." In fact, no human knows what the real laws of nature are, and they will never know[1]; physicists "know" them only as conjectures which are open to refutation.[2] The laws of Nature constrain all men and all activities in the universe.

5.1.1 Since Galileo, the Inventor and Physicist

The modern patent system, in a sense, draws inspiration from Galileo Galilei (1564–1642), the father of modern physics, who in 1594, came up with an invention, a machine for raising water and irrigating land for which he sought a "privilege" (in modern parlance, a patent) on the condition that it had never before been thought of or made by others. In his petition he said, "it not being fit that this invention, which is my own, discovered by me with great labour and expense, be made the common property of everyone." He added that if he were granted the privilege, "I shall the more attentively apply myself to new inventions for universal benefit."[3] His concern: the invention, once divulged, would be copied by others for free exploitation. He, therefore, wanted to reserve some benefits for himself as just compensation for his inventive efforts. The Venetian Council saw merit in Galileo's petition and granted him a "privilege" for 21 years. Since then, his reasoning pervades the patent system.

In 1623, the same Galileo wrote,

> Philosophy [*i.e.* physics] is written in this grand book, the universe, which stands continually open to our gaze, but it cannot be understood unless one first learns to comprehend the

---

[1] Bera (2015).

[2] Popper (1963).

[3] Inkster (2006).

language and interpret the characters in which it is written. It is written in the language of mathematics, and its characters are triangles, circles, and other geometrical figures, without which it is humanly impossible to understand a single word of it; without these, one is wandering around in a dark labyrinth.[4]

In 1637 René Descartes (1596–1650) published his masterwork, *Discourse on the Method of Reasoning Well and Seeking Truth in the Sciences.*[5] In that he unified geometry and algebra for the first time into what we now call coordinate geometry. Descartes invented coordinate geometry by assigning number-pairs to the points of plane Euclidean geometry and proved geometrical theorems about points by proving algebraic theorems about numbers. Euclidean geometry was thus reduced to a branch of algebra. Its remarkable advantage was that one "could borrow all that was best both in geometrical analysis and in algebra and correct all the defects of the one by help of the other." A few centuries later, computer graphics became possible with ease because geometric figures could be equivalently expressed in algebraic form and plotted on a computer screen pixel-by-pixel. Descartes, unintentionally, had enabled the future of modern computer graphics.

While Descartes was alive, in the year Galileo died, an intellectual colossus, Isaac Newton (1642–1727), was born, who put physics on a sound mathematical footing. His book *Philosophiæ Naturalis Principia Mathematica*[6] ("Mathematical Principles of Natural Philosophy"), first published in 1687, laid the foundations for classical mechanics. He also shares credit with Gottfried Leibniz for the development of calculus. For the first time, one could get a feel for the universe in precise mathematical language that described the action of forces on matter and its motion rather than from divinity.

While Newton pinned down the mathematical description of the gravitational force acting between masses, James Clerk Maxwell (1831–1879) nearly two centuries later provided the mathematical description of the electromagnetic force[7] (1864). Newton's equations of motion and his law of gravitation complemented by Maxwell's equations of electromagnetism pretty much made up the *force* and *motion* knowledge required to deal with the engineering and technology of the time. During Maxwell's lifetime, the laws of thermodynamics were also discovered with a central role played by Léonard Sadi Carnot (1824)[8] and Rudolf Clausius (1850).[9] In 1854, Lord Kelvin gave a definition of thermodynamics as follows:

> Thermo-dynamics is the subject of the relation of heat to forces acting between contiguous parts of bodies, and the relation of heat to electrical agency.[10]

---

[4] Galileo (1623). The only modern scientist known by his first name.

[5] Descartes (1637).

[6] Newton (1687).

[7] Maxwell (1865).

[8] Carnot (1824). Carnot introduced the first modern definition of *work* as weight lifted through a height.

[9] Clausius (1850). Clausius defined the term *entropy* as the heat lost or turned into waste.

[10] Thomson (1854). In this paper, William Thomson (Lord Kelvin) first coined the term *thermodynamics*.

The laws of thermodynamics establish fundamental mathematical relations between *work*, *energy*, and *temperature*, along with certain general constraints[11] that are common to all materials. In particular, it established the crucial notion of *entropy*, which essentially is a measure of the number of specific ways in which a thermodynamic system may be arranged. So, this is where the best of scientific knowledge, with precise mathematical descriptions, was when the industrial revolution (1760–1840) ushered in the industrial economy. The French Revolution began in 1789. Notable inventions made during the industrial era include: steam engine (James Watt, 1769; it became a major driver of the industrial revolution); sewing machine (Thomas Saint, 1790); vaccination (Edward Jenner, 1796); the telegraph (Samuel Morse, 1837); rubber vulcanization (Charles Goodyear, 1839); internal combustion engine (Jean Lenoir, 1858); typewriter (1860s); the telephone (Alexander Graham Bell, 1876); the electric bulb (Thomas Alva Edison, 1879); first practical automobile powered by an internal combustion engine (Karl Benz, 1885); AC motor and transformer (Nikola Tesla, 1888); first human-controlled, powered and sustained flight of a heavier-than-air airplane (Wright brothers, 1903); etc.[12] Gradually, the artisan was receding into the background and the STEM professional was coming to the fore. The industrial stage lasted only a few centuries and thus acted as a transitory phase before ushering in the post-industrial stage where we now stand.[13] The transition essentially reflected a fundamental change in the motive power driving economies—from brawn power augmented by industrial machines to brain power augmented by computing machines, and with it the source of innovation—from the artisan to the university educated knowledge professional.

5.1.2 Knowledge Explosion Since the 20th Century

By 1900, scientific knowledge had advanced so much, or so it seemed, that Lord Kelvin (William Thomson, 1824–1907) told an assemblage of physicists at the British Association for the Advancement of Science in 1900, "There is nothing new to be discovered in physics now. All that remains is more and more precise measurement."[14] A similar statement is attributed to the American physicist Albert Michelson made in 1894, "The more important fundamental laws and facts of physical science have all been discovered, and these are now so firmly established that the possibility of their ever being supplanted in consequence of new discoveries is exceedingly remote... Our future discoveries must be looked for in the sixth place of

---

[11] For example, it forbids the existence of a perpetual motion machine in Nature.

[12] "In science credit goes to the man who convinces the world, not the man to whom the idea first occurs." (Francis Galton).

[13] In comparison, the agricultural economy preceding it spanned about 12,000 years. See, e.g., Bernstein (2004).

[14] Cited from: Kelvin, Lord William Thomson (1824–1907). Wolfram Research. http://sciencewo rld.wolfram.com/biography/Kelvin.html.

decimals.".[15] In 1895, Kelvin had confidently said, "heavier-than-air flying machines are impossible" (Australian Institute of Physics), and in 1896 he said, "I have not the smallest molecule of faith in aerial navigation other than ballooning... I would not care to be a member of the Aeronautical Society."[16] How wrong he would prove to be within a few years!

Almost immediately, starting 1900, some breathtaking advances in physics (quantum mechanics (1900–1926), and theory of relativity (1905, 1916)), heavier than air flying machines (Wright brothers, 1903), a deep understanding of mathematics (Gödel's theorem, 1931), mathematical algorithms and computing (the abstract Universal Turing Machine, 1936), the discovery of the structure of the genetic information carrying DNA molecule (the double helix, 1953), the microchip (1958), and men stepping on the surface of the Moon and safely returning to Earth (1969) would be accomplished before 1970. This explosion of knowledge in STEM and its application was phenomenal. By 1934, it had become clear to Karl Popper (and many scientists) that "The game of science is, in principle, without end. He who decides one day that scientific statements do not call for any further test, and that they can be regarded as finally verified, retires from the game."[17]

That game took a new turn in 1936 when Alan Turing, in trying to answer a deep mathematical question, described how one could mechanize the human act of computing. He essentially created an abstract mathematical model (the Universal Turing Machine, UTM) of a human–computer[18] (e.g., a human trained to accurately follow instructions without applying his mind using an unlimited supply of paper, pencil, and erasure—a human robot). It is now well established that Turing machines, recursive functions, λ-definable functions, cellular automata, pointer machines, bouncing billiard balls, Conway's Game of Life, etc. are equivalent in terms of what they can and cannot compute. Thus, the set of computable problems does not depend on the computational model. The abstract UTM thus serves as a generic written description of all classical physical computers.

Then, Claude Shannon in 1948 lucidly provided a mathematical theory of information and connected it with physics (Shannon entropy) and discovered fundamental limits on signal processing operations such as data compression and the reliability of communicating and storing data.[19] In 1953, the remarkable discovery by James Watson and Francis Crick of the double-helix structure of cellular DNA (deoxyribonucleic acid)[20] and that the DNA molecule encodes within it all the genetic information needed to replicate itself[21] turned out to be the biggest discovery in biology

---

[15] Cited from http://www.phy.davidson.edu/FacHome/thg/320_files/physics-is-dead.htm.
[16] Quotes of Kelvin as they appear in: Kelvin, Lord William Thomson (1824–1907), Wolfram Research, http://scienceworld.wolfram.com/biography/Kelvin.html.
[17] Popper (1934).
[18] Turing (1936).
[19] Shannon (1948).
[20] Watson and Crick (1953a).
[21] Watson and Crick (1953b).

since Darwin's theory of evolution (1859).[22] In 1961, Rolf Landauer complemented Shannon's theory with the deep insight that "information is physical" and provided the lower theoretical limit of energy consumption in computation.[23]

These path-breaking events in mid-twentieth century brought about a far greater understanding of nature in mathematical terms than ever before. "Now the language of information is pervasive in molecular biology—genes are linear sequences of bases (like letters of an alphabet) that carry information (like words) for the production of proteins (like sentences). The process of going from DNA sequences to proteins is described by words like "transcription" and "translation", and we talk of passing genetic "information" from one generation to another. It is rather uncanny that molecular biology can be understood by ignoring chemistry and treating the DNA as a computer program (with enough input data included) in stored memory residing in a computer (the cellular machinery). It is this aspect that bioinformatics exploits in deciphering the information carried by the DNA. It is analogous to viewing Euclidean geometry not in terms of drawings but in terms of algebra. In our current understanding, DNA is an informational polymer. It is a vast chemical information database that *inter alia* carries the complete set of instructions for making all the proteins a cell will ever need."[24]

Albert Lehninger lyrically put it: understanding the DNA is the study of "the molecular logic of the living state."[25] Indeed organisms are defined by the information encoded in their genomes. DNA is Nature's digital recording medium. Researchers are now close to anticipating and preempting evolutionary events that left to themselves would perhaps take a few million years to occur, and even of resurrecting extinct species.

The following quotes show the power of mathematics as a descriptive language.

1. "Mathematics is a language plus reasoning; it is like a language plus logic. Mathematics is a tool for reasoning. ... [I]t is impossible to explain honestly the beauties of the laws of nature in a way that people can feel, without their having some deep understanding of mathematics."[26] (Richard Feynman)
2. "The Unreasonable Effectiveness of Mathematics in the Natural Sciences".[27] (Eugene Wigner)
3. "Our reality isn't just described by mathematics – it is mathematics ... Not just aspects of it, but all of it, including you." In other words, "our external physical reality is a mathematical structure".[28] (Max Tegmark)

Mathematicians did not create mathematics with the aim that one day physicists would find it useful. John von Neumann noted:

---

[22] Darwin (1859).
[23] Landauer (1991).
[24] See, e.g., Bera (2015b).
[25] Nelson and Cox (2006).
[26] Feynman (1965).
[27] Wigner (1960).
[28] Tegmark (2014).

> A large part of mathematics which becomes useful developed with absolutely no desire to be useful, and in a situation where nobody could possibly know in what area it would become useful; and there were no general indications that it ever would be so. By and large it is uniformly true in mathematics that there is a time lapse between a mathematical discovery and the moment when it is useful ....[29]

Douglas Hofstadter, in a remarkable book,[30] showed how Gödel's theorem can be understood by analogy with Bach's musical compositions and Escher's paintings thereby showing that even those who revel in the arts can find tremendous beauty in mathematics. The modern computer scientist is not surprised. After all he encodes music and paintings in abstract binary strings (just as easily as he encodes mathematical algorithms) which a computer (using appropriate software and input–output hardware) decodes into music and painting at will.

One might thus conclude that the Universe itself is a computer ceaselessly performing mathematical calculations of the laws of Nature. We have traversed far from when Galileo famously said that the universe is written in the language of mathematics to Max Tegmark saying that the universe IS mathematics. If Tegmark's conjecture is right (there is no convincing refutation of it yet), his thesis represents a paradigm shift in the relationship between physics and mathematics. *Ipso facto*, it fundamentally affects how we define patent-eligible subject matter; an invention's utility, novelty and non-obviousness; how an invention is described; and the expansive scope of the doctrine of equivalents in patent law.

The twentieth century began by dazzling us with aeroplanes, automobiles, and radio and ended with spaceships, computers, cell phones, the Internet, and genetic engineering.[31] In just a century they dramatically changed the industrial economy into a post-industrial economy that is global, heavily consumer-oriented, talent-hungry, knowledge-centered, dependent on a university-educated and globally mobile workforce, and above all driven by innovation as never before.[32] We are at the knee of an exponentially rising curve, advancing at an exponential rate in developing new technologies, very much in line with Kurzweil's predictions. The STEM knowledge required to keep pace with this rate of advancement is beyond the intellectual capacity of most people armed with a PhD in STEM. The time to start reforming the patent system was when Neil Armstrong stepped on the surface of the Moon! We are more than half-a-century behind schedule.

---

[29] Neumann (1954).

[30] Hofstadter (1979).

[31] For a timeline of inventions, see, e.g., http://inventors.about.com/od/timelines/a/twentieth.htm. See also: Olson (2015) and Cooke and Hilton (2015). "The past half-century has witnessed a dramatic increase in the scale and complexity of scientific research. The growing scale of science has been accompanied by a shift toward collaborative research". "The size of authoring teams has expanded as individual scientists, funders, and universities have sought to increase research productivity and investigate multifaceted problems by engaging more individuals. Most articles are now written by 6–10 individuals from more than one institution".

[32] See, e.g., Palmisano (2003). See also: Bera (2015a).

5.1.3 Algorithmically Designed Biological Inventions

In 1973, the pioneering work of Cohen and Boyer in recombinant DNA technology[33] gave birth to genetic engineering and the biotechnology industry. The related Cohen–Boyer patents (U.S. Patent Nos. 4,237,224; 4,468,464; and 4,740,470) that protected the technology played a stellar role in the rapid rise of the biotechnology industry.[34] The next landmark was the creation of a bacterial cell controlled by a chemically synthesized genome by Craig Venter and his group in 2010.[35] Then in 2014, Floyd Romesberg and colleagues[36] reported the creation of a semisynthetic organism with an expanded genetic alphabet. The new letters in the alphabet were artificially created nucleotides not found in Nature. Along with these breakthroughs, the great promise of CRISPR (clustered regularly interspaced short palindromic repeats), and in particular CRISPR-Cas9 gene editing technology pioneered by Feng Zhang,[37] Emmanuelle Charpentier, and Jennifer Doudna[38] in 2012 as a new way of making precise, targeted changes to the genome of a cell or an organism has set the stage for major advances in synthetic biology. The achievable aim is to design and construct new biological parts, novel artificial biological pathways, organisms or devices and systems including the re-design of existing natural biological systems for useful purposes. Thus, the focus is on developing tools and methods that would enable researchers to encode, in artificially created or natural DNA, basic genetic functions in novel combinations by design. The aim is to artificially create biological systems of increasing size, complexity, and tailored functionality. Currently, synthesis capabilities far exceed design capabilities in the sense that we know how to build but not yet with clarity what to build.[39] Synthesis capabilities have developed to a state where DNA synthesis can be automated, and the desired DNA produced once the sequence is provided to vendors.

Biologists now have tools for manipulating DNA in a manner similar to manipulating character strings in a text. For example, they can copy DNA fragments using the polymerase chain reaction (PCR) or clone it using a cloning vector; cut DNA using molecular scissors called restriction enzymes; join two complementary DNA strands into a double-stranded molecule in a process called hybridization; and measure the size of DNA fragments without sequencing them using a technique called gel-electrophoresis. This complex integration of biology and traditional engineering driven by information processing and computing technologies, and algorithm design

---

[33] Cohen et al. (1973).

[34] See, e.g., Bera (2009, 2012).

[35] Gibson et al. (2010).

[36] Malyshev et al. (2014).

[37] Cong et al. (2013).

[38] Sharlach (2014). The Nobel Prize in Chemistry 2020 was awarded jointly to Emmanuelle Charpentier and Jennifer A. Doudna "for the development of a method for genome editing." For a description of CRISPR-Cas9, see their Nobel Lectures at (Charpentier) https://www.youtube.com/watch?v=3POrtQEpV2s, and (Doudna) https://www.youtube.com/watch?v=KSrSIErIxMQ.

[39] Prather (2010).

is moving so rapidly that a couple of decades hence, researchers may begin producing synthetic organisms designed to produce not only pharmaceutical products but also industrial products such as biofuels on a commercial scale. Possible socio-economic benefits from synthetic biology research are thus enormous.

The CRISPR genome editing technology allows one to precisely insert DNA into a cell in vivo or snip out mutated DNA and replace it with the correct sequence. It thus offers possible means of treating many genetic disorders.[40] The key to success will lie in encoding and embedding mathematically structured and manipulated information in DNA strings and in interpreting it or activating it in a given chemical context. A 2012 book, *Fueling Innovation and Discovery: The Mathematical Sciences in the 21st Century*, from the National Academies explains how mathematics is fueling innovation and discovery. It notes,

> The mathematical sciences are part of everyday life. Modern communication, transportation, science, engineering, technology, medicine, manufacturing, security, and finance all depend on the mathematical sciences, which consist of mathematics, statistics, operations research, and theoretical computer science. In addition, there are very mathematical people working in theoretical areas of most fields of science and engineering who also contribute to the mathematical sciences. There is a healthy continuum between research in the mathematical sciences, which may or may not be pursued with an application in mind, and the range of applications to which mathematical science advances contribute. To function well in a technologically advanced society, every educated person should be familiar with multiple aspects of the mathematical sciences.[41]

Creating public awareness about mathematical sciences in the post-industrial economy is now a critically felt need. It is widely believed that the twenty-first century belongs to advances in life sciences. The century has already begun creatively by creating and editing novel and non-obvious DNA sequences which speak for and describe themselves in a language that those skilled in the art understand with precision, as to the invention the sequences stand for. These are self-describing inventions just as mathematical algorithms are when interpreted in a well-defined context.

5.1.4 Molecular Biology is Mathematical

We define the information content of an object as the size of the set of instructions that we need to be able to reconstruct the object, or better, the state of the object. Implicit here is that information can be encoded in physical systems. Indeed, without a physical device we cannot store, transmit, process, or receive information. Moreover, the laws of physics dictate the properties of these devices and therefore they limit our capabilities for information processing. Hence, it is clear that information theory cannot be a purely mathematical concept but that the laws of physics dictate the properties of its basic units. This rather obvious fact became obvious to information

---

[40] Yin et al. (2014).
[41] NRC (2012).

theorists only in 1961 with the publication of a landmark paper by Rolf Landauer,[42] who realizing that physical devices are needed to encode information, showed that there is a fundamental asymmetry in the way Nature allows us to process information. In fact, he proved the surprising result that all but one operation required in computation could be performed in a reversible manner. For example, copying classical information can be done reversibly and without wasting any energy, but when information is erased there is a minimum energy cost involved per classical bit to be paid. That is, the erasure of information is inevitably accompanied by the generation of heat (i.e., there is friction and resistance and creation of randomness). Indeed, Landauer's principle provides a bridge between information theory and physics via thermodynamics. That insight has brought about a sea change in the way we look at information and computation as the following quote from the quantum physicist David Deutsch shows:

> The theory of computation has traditionally been studied almost entirely in the abstract, as a topic in pure mathematics. This is to miss the point of it. Computers are physical objects, and computations are physical processes. What computers can or cannot compute is determined by the laws of physics alone, and not by pure mathematics. One of the most important concepts of the theory of computation is *universality*. A *universal computer* is usually defined as an abstract machine that can mimic the computations of any other abstract machine in a certain well-defined class. However, the significance of universality lies in the fact that universal computers, or at least good approximations to them, can actually be built, and can be used to compute not just each other's but the behavior of interesting physical and abstract entities.[43]

Modern biologists view the deoxyribonucleic acid (DNA) as a string of encoded information, something like a long tape containing both program and data for a Universal Turing Machine. The DNA's interaction with the rest of the cell's machinery is nothing but a series of computational steps. Not surprisingly, bioinformatics as a discipline has so many computer scientists in its ranks, many of them holding joint academic appointments in the departments of biology and of computer science.

Molecular-biology-rooted biotechnology inventions are expected to dominate 21st-century commerce because of biotechnology's tremendous potential to contribute to human health, food security, and the environment in which humans live. These inventions are clearly patentable subject matter. All players involved in creating and commercializing this knowledge-and-capital intensive emerging technology are obviously deeply interested in knowing how they would gain or lose from the intellectual property (IP) system in place and whether that system needs to be changed, replaced, or abolished from their respective perspective.[44] The patent system must address their concerns in a comprehensive way as to subject matter eligibility in patent law.[45]

---

[42] Landauer (1961, 1991).
[43] Deutsch (1998), p. 98.
[44] Bera (2015b).
[45] See, e.g., Bera (2016), Bera (2015c), and Bera (2015).

5.2 Quantum Physics

Quantum mechanics, the most successful branch of scientific knowledge, deals with the world inhabited by photons, electrons, protons, atoms, molecules, etc. and how they interact among themselves to create larger matter entities in terms of chemical bonds of various strengths. It does so using the abstract language of mathematics, and it is only in that language that we understand it.

> Quantum mechanics is an immensely successful theory. Not only have all its predictions been experimentally confirmed to an unprecedented level of accuracy, allowing for a detailed understanding of the atomic and subatomic aspects of matter; the theory also lies at the heart of many of the technological advances shaping modern society – not least the transistor and therefore all of the electronic equipment that surrounds us.[46]

Quantum mechanics is expected to play an important role in synthetic biology, inter alia, in understanding the myriad chemical reactions that take place inside a cell and the chemical means a cell uses in information transfer within itself and the external world. The patent system is not geared to deal with a flood of quantum mechanics-based inventions.

We now have reasons to believe that these "biological" computations are not completely based on classical logic but also on quantum logic. It turns out that quantum computers can do what classical computers do plus some more.[47] The surprises and breakthroughs will most likely come from the exclusively quantum part of the logic. It is a realm of logic where our normal human reasoning fails. It is a branch of knowledge, which is understood with difficulty even by the experts in the field! No one has yet claimed to have developed any intuition for it. The new fields of quantum computing and quantum information have already accumulated some breathtaking results in teleportation, encryption and code breaking, parallel computing, etc. What disruptive technologies they will spawn are difficult to foretell. That there will be more stunning results coming out of quantum mechanics with technological implications seems obvious. That it will eventually encompass biology, as it did astronomy some decades ago, appears inevitable. Interestingly, the abstract mathematical representation of quantum mechanics captures several interpretations of the universe.[48]

---

[46] Zinkernagel (2015) and Zinkernagel (2016).
[47] Nielsen and Chuang (2000).
[48] See, e.g., Bera and Menon (2009).

5.3 Confluence of AI, Synthetic Biology, and Quantum Computing

Patent systems around the world were designed to deal with inventions created in siloed disciplines. Thus there are mechanical, electrical, chemical, biological, etc. inventions dealt by patent examiners specialized in these disciplines. Today, the most interesting examples come from teams that juxtapose concepts from multiple disciplines and the most advanced teams include elements of AI into their work. Their inventions are more easily understood in conceptual terms which emphasize an invention's functionalities rather than in terms of their physical representation. This is increasingly true where the functionalities are carried out by embedded computer chips running algorithms to provide functionalities. When abstract concepts are captured in mathematics, mathematics is captured in software, and software is hard-wired in computer chips or some other material form, how the invention is captured becomes irrelevant because the forms are interchangeable. Abstract concepts and their material representation are two sides of the same coin; they define each other. If one side is patentable, then the other side must be too. Abstraction cannot be treated differently from material representation. Further, by invoking the doctrine of equivalents a vast number of inventions can be nullified, e.g., by citing an old, outdated patent and tying it to an abstract concept to claim that the concept covers all conceivable form of its material representation and hence any later patent that can be tied to that concept either belongs to prior art or is infringing a prior patent. Is this ridiculous? No, because future AI machines can make this claim using a human intermediary! The relationship between a PHOSITA and prior art is that a PHOSITA can always improve his knowledge and application of that knowledge by making a diligent study of the prior art. A human PHOSITA in today's competitive world should be assumed to be a person who is alert, inquisitive, and willing to imbibe the prior art if called upon to do so in solving a problem. In an increasing number of situations, AI machines are being programmed to do so; the AI machine is the PHOSITA.

Thus, we are forced to redefine what is patentable subject matter, the relevance of granting patents to human inventors, the criteria to be used to define the legally enforceable boundary of a patent to detect infringement, and above all who is a PHOSITA. Without knowing who the PHOSITA is related to a patent application, it is no longer possible to even decide if the patent applicant is an inventor.

5.3.1 Knowledge Integration by Concepts

Computing technology has now advanced to a stage where quantum computers can mimic a Universal Turing Machine (UTM) and beyond. A quantum computer's phenomenal computing power comes from the extraordinary laws of quantum mechanics that include such esoteric concepts as superposition of quantum states, entanglement ('spooky action at a distance' as Albert Einstein once quipped) and tunneling through insulating walls, which though highly counter-intuitive, play extremely useful roles in understanding Nature at subatomic levels. It appears that these concepts cannot be ignored in biology and living processes in the way they are ignored, say, in the design of cars and airplanes. There are areas in biology where quantum effects have been found, e.g., in protein–pigment (or ligand) complex systems.[49] Thus, while the role of quantum mechanics is clear in quantum computing and hence in advancing both AI and synthetic biology research, it is not yet known if in the design of DNA, knowledge of quantum mechanics is required or that natural selection favors quantum-optimized processes. Essentially, we do not know if any cellular DNA maintains or can maintain sustained entangled quantum states between different parts of the DNA (even if it involves only atoms in a nucleotide). But we cannot rule out the possibility that sporadic random entanglements do occur that result in biological mutations or that researchers will not be able to achieve it in the laboratory and find novel uses for it in synthetic biology.[50] For example, in principle, it is possible to design molecular quantum computers, insert them in cells that can observe cellular activity, and activate select chemical pathways in the cell in a programmed manner. There is increasing speculation that some brain activity, e.g., cognition, may be quantum mechanical.[51]

The role of computer simulation in developing modern inventions is barely understood by patent examiners, mainly because it rests on mathematics, its digitization for calculation, and inspired interpretation of mathematical models in different branches of science, engineering, and technology.

The fundamental role of mathematical simulation is to capture the abstract essence of algorithmic changes that define a system's dynamics or structure without attaching meanings to it. In mathematics symbols have no meaning other than those implied by their relationships to one another. It therefore lends itself to automation via the isomorphism shown in Table 3.2, Chap. 3. It is the most powerful means by which we generalize, that is, we identify the parts of a whole, as belonging to a much larger whole at a conceptual level. This means that the expansive use of the doctrine of equivalents from the standpoint of concepts can create havoc by (1) allowing a patentee to claim an ever-expanding scope for his claims, and (2) an alleged infringer to counterclaim that the patent in suit is invalid because it belongs to the prior art as it falls into the expanded scope of one or more earlier patents that fall in the same conceptual category. Conceptual bases is what well-qualified STEM researchers (the modern PHOSITA) routinely use in simulation, e.g., studying mechanical systems by studying their exact analogous electrical systems because they share a common mathematical model. Since mathematics is mechanizable (Turing 1937),[52] and AI wholly depends on mathematics and computers to execute it without any application

---

[49] Brookes (2017).
[50] Brooks (2015).
[51] Fisher (2015).
[52] Turing (1936).

of the mind, the doctrine of equivalents by itself would be enough to completely demolish the patent system of any country.

5.3.2 Integrating the Triad: Mechanization of Speciation

A combination of emerging technologies such as clustered regularly interspaced short palindromic repeats (CRISPR), artificial intelligence (AI), and quantum computing (QC); new delivery models for products and services that form the core around which *Homo sapiens* organize themselves through collaborative division of labor; and talent migration, driven not by rote education but by innate creativity and global opportunities for employment open to them, is disrupting and changing the character of the global talent pool that society needs today. Globalization has created opportunities for the talented to reach for the skies but in a resource constrained world it also means that many others must be or feel deprived.[53]

This social dynamics is captured very well in terms of a remarkable result in graph theory[54] and the logistic map in chaos theory[55] because the related mathematical models are equally valid for both animate and inanimate systems. The models show that in a resource constrained world very rapid progress provides ample scope for swift and fluctuating adversarial social dynamics to occur in which some turn predators and others become preys depending on current circumstance. Globalization has accentuated the problem at all levels of social structure, and since speciation is triggered by a changing environment, it affects the DNA. This has imposed survivability demands on the *Homo sapiens*.

As this pressure mounts beyond endurance, *Homo sapiens* will face speciation by natural selection with uncertain outcomes. However, in the case of *Homo sapiens* in their present state of STEM knowledge, this process too may face a disruptive change because the highly intelligent among them may boldly initiate speciation using upcoming advances in synthetic biology, perhaps after perfecting their techniques by creating humanoids (a hybrid creation of life with embedded intelligent machinery). This will be a watershed event where a species takes on the task of speciation on itself.

This remarkable possibility arises because *Homo sapiens* created and mastered mathematics, rational thought, computing machinery, and eventually deep data analytics so that life could be designed by them in the laboratory to create superior species. This will also permit us, at all levels of the hierarchy of biological structures (molecules, cells, tissues and organisms), to redesign existing natural biological systems and may even help us recreate certain extinct species (if we can also recreate the environment they had adapted to). It is not surprising that an extinct species has never revived itself since speciation and environment go together. Successes of

---

[53] Bera (2019).
[54] Erdős and Rényi (1960).
[55] May (1976).

synthetic biology will change the face of human civilization and almost certainly bring in new elements into play when *Homo sapiens* eventually speciate by playing an active role in it. The present patent system cannot control inventions arising from this development.

Since the discovery of the double helix structure of cellular DNA by James Watson and Francis Crick in 1953[56] and its significance that the "precise sequence of the bases is the *code* which carries the genetical *information* ..." (emphasis added),[57] the jargon and theory of information has invaded molecular biology. This enriched biotechnology and computational biology with nomenclature, definitions, concepts and meanings, which facilitates integration of synthetic biology with AI and QC. DNA is an information carrying polymer. It is an organized chemical information database that *inter alia* carries the complete set of instructions for making all the proteins a cell will ever need.[58]

DNA synthesis services are now commercially available. The time is ripe for the wholesale integration of synthetic biology with AI via mathematics to enable seamless communication among them, connect with and discover conceptual similarities for consistent integration of subsystems and validation of the whole system. The added benefit is that it can be used to also communicate between humans and machines. It is fortuitous that the DNA serves as the "Book of Life" that appears to have structure and grammar amenable to translation into mathematics. Once translated, biologists will discover some amazing patterns that have a direct bearing on life at the molecular level.

5.4 When Machines Invent, Are Patents Relevant?

Patent law allows useful, novel, non-obvious, and well-articulated inventions to be patented subject to certain other statutory requirements being fulfilled, before the inventor is given a limited period property-like monopoly ownership of the invention. The principal feature of this ownership is the statutory right to sue those who infringe the patent, but not necessarily the right to practice the invention. The heart of any patent is its set of claims which delineate the boundary of the protected territory of the invention's novel and non-obvious aspects with respect to the related technological prior art as statutorily defined. Claims are required to fulfill two important functions: (1) to give notice to the examiner at the U.S. Patent and Trademark Office (USPTO) during prosecution as to what is being claimed for limited period monopoly privileges, and (2) to the public at large, including potential competitors, after the patent has issued, what is not to be infringed during the term of the patent. Once the patent expires, it is dedicated to the public.

---

[56] Watson and Crick (1953a).
[57] Watson and Crick (1953b).
[58] For a more elaborate explanation see, e.g., Bera (2019).

The Supreme Court of the United States (SCOTUS) has held that claims define the scope of patent protection: "[T]he claims made in the patent are the sole measure of the grant …".[59] The courts also hold that 'subject matter disclosed but not claimed in a patent application is dedicated to the public'.[60] In reality, most patent claims, due to inadequacies of natural languages in which they are written, and inadvertent omissions and inclusions during drafting become prone to contentious ambiguity as to the scope of the claims. Later, if the patent's validity is contested, failure to establish the boundary would make the claim at issue unpatentable, invalid, and unenforceable under the doctrine of indefiniteness.[61] While literal infringement of a patent is usually obvious and easily dealt with, the grant of a patent also gives the invention additional protection from the judicially created doctrine of equivalents which serves to expand patent protection beyond the literal language of the claim. That is,

> A patentee may invoke this doctrine to proceed against the producer of a device "if it performs substantially the same function in substantially the same way to obtain the same result."[62]

This immediately raises two questions: (1) "What if a device performs substantially the same function in a substantially different way to obtain the same result?" And, (2), "What happens, if during patent prosecution, the inventor surrenders or dilutes some of his claims, can he reclaim them once the patent is granted under the doctrine of equivalents?" Mathematicians resolve such complex issues by ensuring that their language does not allow such ambiguities to occur in the first place. We shall see that this fact has major implications in the development of AI and its introduction in any discipline of knowledge. Presently courts deal with such questions in ambiguous and highly unsatisfactory and tortuous ways, e.g., the reverse doctrine of equivalents[63] to deal with the first question and prosecution history estoppel[64] to deal with the second. The fact is that the manner and the language in which claims are written compounded by the fact that judges are not trained in STEM precludes a satisfactory solution to be ever developed. The most damaging aspect of the situation is that the doctrine of equivalents de facto redefines the scope of the claims and ignores the public's right to know in advance the precise legal limits of patent protection without recourse to judicial ruling. By redefining the scope, moving boundaries of ownership get created where knowing what was and who infringed becomes unusually complicated to the extent that chaos may prevail making a judicial decision impossible. The SCOTUS, if it anticipates such a situation, can, of course, wriggle out by not hearing the case at all without assigning any reason! In many other cases, either ignorance or uninformed heroism leads judges to resort to Markman hearing[65]

---

[59] SCOTUS (1961)
[60] See, e.g., CAFC (2002).
[61] SCOTUS (2014).
[62] SCOTUS (1950).
[63] SCOTUS (1950).
[64] SCOTUS (2002).
[65] SCOTUS (1996).

5.4 When Machines Invent, Are Patents Relevant?

where the interpretation of claims are handled by the judge while questions of validity are handled by the jury. Unfortunately, both lack the STEM expertise required for their respective jobs. These tasks really belong to the Patent Validation Board (see Sect. 5.5.1).

Any mathematician would see that the questions are ill-posed in the patent system in which it is raised. The problem is with the language in which the Patent Act and patent applications are written. The present style and legalese used in claim writing hinders the smooth flow of thoughts. It needs a change. Just as arithmetic (indeed any branch of mathematics) cannot be done comfortably using Roman numerals, mathematicians came up with Arabic numerals and augmented it with other symbols and a set of precise, formal rules for manipulating those symbols to produce permissible words, sentences, etc. so that there is no ambiguity about the legitimacy of the mathematical structures that are built and the way or the multiple ways they are understood. The U.S. courts have further vitiated the already troubled waters by functioning both as courts of law and courts of equity. Oil and water don't mix well.

Further, the patent system failed to note what Harding had already observed in 1941:

> Originally industry relied on the chance discoveries of gifted individuals working at random, their choice of problems being guided by their interests, backgrounds, abilities and the prospect they saw of making a profit from their activities. Modern research is planned to fit specific needs. A large element of unpredictability and discovery and in the value of discoveries in monetary terms, can no longer be permitted. [The so-called discovery and invention of serendipity.] In the 20th century industry saw that it could no longer rely on random discoveries and it turned to the accumulation of new knowledge. The science of invention was perfected and research discoveries were largely tailored to specific business or industrial requirements.[66]

The need for and reliance on patents has dramatically changed over time. In such circumstances, a court of law must tread with caution when injecting issues related to equity in its judgements. For example, unclaimed subject matter in patent claims should strictly be considered as dedicated to the public and not surreptitiously reclaimed under the doctrine of equivalents or equity considerations.

As STEM knowledge advances, it increases the possibility that certain inventions are quite likely to sprout spontaneously and hence such inventions, if patented, only hinders the public's access to such inventions. The essence of the patent system is to strike a balance between private incentives and protection of public interest, not indiscriminate distribution of private incentives. It is therefore necessary to assess if the invention may have arisen without the incentives.

All these issues become irrelevant when an AI machine invents. In reality, an AI machine does not invent, it merely reproduces "theorems" in a mechanized way that are implicit in the axiomatic system that is programmed into it or it becomes a means of generating axiomatic systems. All inventions (theorems) produced using AI reside in an abstract world from which innumerable specific instances to satisfy present patent acts in the world can be created. This means the PHOSITA has no place

---

[66] Harding (1941).

in an AI-driven world. Any AI machine is a PHOSITA and an inventor. AI-machines can churn out inventions on demand and spontaneously.

> The way in which the building blocks of a body of thought are designated profoundly affects the development of that discipline. — H. C. von Baeyer, a noted physicist.

AI is now at a stage where it influences the way we solve problems, especially, how we solve problems rationally. Logic is the foundation of rational human thought. It deals with the terms "and", "or", "not", "if", "then". Reasoning (or propositional calculus) is built around our notions of the correct usage of the words *if … then …* (or *implies*), *or*, *and*, *not*. It has a vocabulary, rules that tell us how to construct correctly formatted statements, and inference rules for deriving new statements from a given set of correctly formatted statements. The inference rules are chosen such that if the statements in a given set represent true statements, then subsequently derived statements will also represent true statements.

Logic underpins mathematics, and the natural sciences, especially, physics. The great advances in mathematics and the sciences were made possible because mathematicians, in particular, meticulously developed a symbolic system to express their concepts, axioms, theorems, and proofs. When physicists adopted mathematics as their lingua franca, it began to advance rapidly, as have chemistry, biology, and engineering since. With the publication of Isaac Newton's *Philosophiae Naturalis Principia Mathematica* in 1686, scientists and later engineers have gone from strength to strength using the power of mathematics. Jurisprudence, although founded on logic, took a different route; it continued to rely on natural languages (with its built-in ambiguities) for communication. Thus, courts cannot always interpret the law literally but must try to divine the intent behind the law. Literalist interpreters see it as subverting the statutes. When ambiguity rules, decisions end up as 5–4, leaving a feeling that it may well have been decided by tossing a coin. In interpreting mathematical rules or laws of nature, intent is irrelevant.

Modern technologies have deep roots in science, and science has deep roots in mathematics. The immense power of AI, tapped and untapped, lies in the expressive power of mathematics and the computing power of computers which has yet to reach its zenith. Language powers intelligence.

So there are unavoidable impediments in interpreting the existing patent law. As Benjamin Whorf said, "Language shapes the way we think, and determines what we can think about." And Ludwig Wittgenstein noted, "The limits of my language mean the limits of my world."[67] Nevertheless, the law requires that it must be clear from the written description that the applicant was indeed in possession of the claimed invention at the time of filing but it does not provide unambiguous means to assert what exactly is being claimed.

---

[67] The original statement was in German ("Die Grenzen meiner Sprache bedeuten die Grenzen meiner Welt.").

5.4.1 Fundamental Tests of Patentability

The base reference for measuring human ingenuity is obviously a world in which no humans exist. Therefore the very first test of patentable subject matter is: Could the invention under consideration possibly have occurred or likely to occur in our planet in the absence of intelligent and thinking humans? (*Yes means not patentable.*) For example, it is inconceivable that a modern jetliner could have ever occurred in the absence of intelligent and thinking humans. The second test is: If the invention could occur in the future in our planet devoid of humans, would the presence of intelligent and thinking humans accelerate the process of bringing forth that invention not by their mere presence but by observation, analysis, and deliberate human intervention. (*No means not patentable.*) The third test is that if the problem the invention solves was posed to other humans would several of them (in a statistical sense) have come up with the invention or a similar invention or a superior invention, say, within a specified 'short' period of each other (*Yes means not patentable.*). Finally, if a computer, such as IBM's *Watson*, was given the task of solving the problem, would it solve the problem in a few years (*Yes means not patentable.*). Note that a mathematical solution to the problem is a valid solution. An algorithmic solution is, by definition, implementable on a Universal Turing Machine (UTM) and hence on a sufficiently powerful physical computer. The invention is not patentable if the answer is *not patentable* to any of the four questions.

It is a genetically coded property of the human mind that when it is inquisitive about the material world, it frames, tests, and revises hypotheses (conjectures and refutations) about patterns and correlations that fit what it observes along with the knowledge and information it possesses about the world. When it finds satisfactory patterns or correlations, whether deterministic or statistical, it may use them to solve problems. When the solution leads to a material product, or a process or a correlation that can be executed or determined that includes using non-mental tools or machinery and which necessarily requires in some part human ingenuity then that product or process or correlation is patent eligible subject matter provided its patenting does not unreasonably interfere with or discourage developments or the further spread of useful knowledge itself (the constitutional requirement). This is a subjective decision to be generally decided by majority voting by a statutory body comprising members from the patent office, the national science academies, and other eminent STEM experts, since a bright line rule cannot be formulated. This statutory body therefore will have the enormous moral responsibility of deciding when the grant of a patent to a given invention would adversely affect the society at large based on parameters that cannot be objectively quantified. That is, this august body must deal with "the difficulty of drawing a line between the things which are worth to the public the embarrassment of an exclusive patent, and those which are not".[68] Then and only then should the grant of a patent be decided on the basis of the patent act. It is at this stage that the USPTO has the great responsibility of ensuring that it does not grant overbroad and indefinite claims, which once admitted, would invite litigation.

---

[68] Jefferson (1813).

5.4.2 The Superfluous PHOSITA

When dealing with the doctrine of equivalents, it is a fallacy to bring in the PHOSITA into the analysis for infringement. Those who want to circumvent a valid patent and succeed must be deemed to be superior to a PHOSITA and in the class of inventors capable of conjuring patentable inventions. If a patent-in-suit is valid it is an advancement in technology by definition. Therefore circumscribing it would likely require another advancement. This is where modern researchers well-versed in mathematics are likely intermediaries who can find equivalents based on conceptual similarities rather than in terms of physical similarities. In engineering this is called computer simulation where, say, mechanical artifacts can be analyzed in digital simulation based entirely on mathematical calculations and reinterpreted into an equivalent electrical artifact which is another discipline in engineering. With 3D manufacturing and CAD/CAM technologies easily available and continuously advancing with embedded AI, an extremely wide range of granted patents may be assiduously converted into equivalent non-inventions, more-or-less, in a mechanizable way. No judiciary in the world has enough knowledge to deal with this situation, nor does any patent office in the world.

Although Albert Einstein failed in his quest to find a theory of everything, it does not follow that an AI machine will not find or make considerable advances toward it. Once it does, then a conceptual solution to a problem would more-or-less amount to finding equivalent solutions in a wide-range of engineering disciplines. We are approaching this stage rather rapidly, going by Kurzweil's forecast.

5.4.3 Drafting of Patent Applications

All patent systems around the world suffer from serious debilities. The first and foremost is its gross inability to move away from archaic legalese in drafting patent applications to describe an invention. It has shown no sign of rectifying this. This anachronistic system bears comparison with arithmetic which very sensibly switched over from using Roman numerals to Arabic numerals. Without the switch, human civilization would still be in the preindustrial era. With Roman numerals, even adding a grocery bill is an onerous and frustrating task. The primacy of language seems to have escaped the notice of the IP community in granting patents and in adjudicating disputes related to them.

A defining feature of a sophisticated society is how it communicates with humans, machines, and institutions. That is how humans control and coordinate strategy. But the relationship between language and power is intricate. Thoughts get communicated through language (speech, script, and sign) and emotion (body language). Thus, whoever speaks depends on language, but ultimately the power of language lies

5.4 When Machines Invent, Are Patents Relevant?

not with the speaker but with language itself.[69] Anyone can acquire the power of language, even AI machines.

The frustrating aspect of legalese in the patent system is that even judges must resort to Markman hearings (a common practice since the U.S. Supreme Court, in the 1996 case of *Markman v. Westview Instruments, Inc.*, 517 U.S. 370 (1996)) so that the patent can be understood in plainer English. One can imagine the fate of the jury (comprising mostly of ordinary people with very little or no education in STEM). That such a jury should even be involved in deciding the fate of inventions involving cutting edge technologies in patent trials is irrational. That both judge and jury are ignorant of STEM makes the entire set-up a mockery of rational reasoning. It provides ample opportunities for patent lawyers to game the system in a reckless manner.

The overhaul of the patent system must begin at its very foundations. While doing so one must bear in mind that AI machines are already discovering novel and non-obvious inventions that do not belong to the prior art. Google's alpha-Go is a harbinger of such events occurring more frequently in the future. The AI-machine as a prolific source of inventions is no longer a fond dream but a reality so extraordinary that every AI machine can be viewed as a PHOSITA that far surpasses highly intelligent *Homo sapiens*. The collapse of the patent system is near.

A deep question arises. If a machine reinvents and without conscious thought becomes instrumental in commercializing a patented invention during its term, taking directions from software (essentially mathematical algorithms), unsupervised by humans, does it qualify as infringement or should the patented invention be annulled on the grounds that machine implementable prior art existed which was overlooked by overworked patent examiners. If inventions can sprout from machines for the asking, then what and where is the need to protect, celebrate, and sanctify the rare individual who once in a while measures up to the inventive capability of a machine in terms of ingenuity, novelty, non-obviousness? Further, a fully documented disclosure of the invention always resides in the machine when the invention is in use which it can also share with other machines and human experts. Indeed, at any time we may not even know if it has already disclosed the invention to other machines and so has placed it in the public domain.

In Sect. 5.1 it was noted that the judiciary erred in the past when it declared that the "Laws of Nature" as discovered by humans are unpatentable because they are common property of all mankind.[70] However, it is prudent to deny patents to such conjectured "Laws of Nature" because otherwise the inventor will gain excessive monopoly rights that can be coercively used to stifle others from inventing further

---

[69] Weiß and Schwietring (2018). See also: Bera (2018).
[70] Bera (2016), Bera (2015c).

during the term of the patent. James Watt[71] and Rifkin[72] are notable examples of such actions, while the Cohen–Boyer patent is not.[73] The emerging situation is that AI machines work only according to the unknown laws of Nature that appear to be mathematically encodable in abstract form. The machines do not use any intelligence in the process; they only follow instructions. Under patent law, such machines cannot even apply for a patent since they are not humans. Thus all machine created inventions must automatically fall into the public domain unless protected as trade secrets.

We must also recognize that AI-driven, 3D printers and personal robots producing goods and services on demand in a customized manner will soon become ubiquitous in homes. Such machines can hardly be declared as infringing an active patent. Punitive action against hundreds of millions is impossible. Self-learning AI machines are advancing their ability to solve problems and invent based purely on mechanized implementation of conjectures programmed into them. In fact, they can even be programmed to make conjectures of their own. Eventually, the *Homo sapiens* will no longer be looking for a better patent system because they would be speciating themselves into extinction, possibly within a century. There is a new successor species waiting in the wings to displace us.

5.5 Recommendations for the Immediate Future

In the interim, while society is reorganizing itself to cope with AI, the courts should carefully consider the embarrassment of creating a judicial exception when deciding a patent case. A pragmatic two stage process would be to ask: (1) Is the invention patentable under the statutes? And in the process, pointing out anomalies in the statutes if a reasoned conclusion cannot be drawn. (2) If the invention is patentable, should it be denied a patent because it would be an "embarrassment" to society in its present state and rate of evolution or should it be granted a patent with obligatory social conditions (decided on a case-by-case basis) attached in consultation with the patentee? This question should be answered by a separate statutory body, e.g., PVB, and the answer should not be contestable in a court or any other forum.

The following should be taken into account when considering patent grant:

---

[71] Ashton (1955). James Watt deliberately refused to license certain steam engine related patents held by him to prevent others from bringing improvements into the market and compete against him. He actively discouraged use of steam at high pressure even though it was not covered by his patents. The authority he wielded at the time was sufficient to clog engineering enterprise for more than a generation.

[72] Heathcotte and Robert (2006). Stuart Newman and Jeremy Rifkin unsuccessfully sought a US patent in which they claimed a method for combining human and animal embryo cells to produce a single embryo, which could then be implanted in a human or animal surrogate mother, resulting in the birth of a "chimera". Their unusual objective was to secure the patent and then restrict the application of this technology for the life of the patent, during which they hoped to foster a social debate about moral boundaries in relation to biotechnology patents.

[73] Cohen et al. (1973). See also: Bera (2009) for the story of the magnanimity with which the patents were licensed.

5.5 Recommendations for the Immediate Future

1. *Human ingenuity* criterion. The patent application must describe an invention that required human ingenuity for its creation, without which it is extremely unlikely that the invention would have occurred or might occur at some distant time in the future or occur spontaneously over which humans have no direct or indirect control on planet Earth.
2. *Quid pro quo* criterion. Establish a clearer equitable *quid pro quo* criterion for patent grant. This should be the first barrier a patent application must cross before provisions of the patent act are applied. An invention is patent ineligible if it fails to meet an established *quid pro quo* criterion for patent grant. Any equitable criterion must bear in mind its effect on public health and safety, effect on human skill development, the functioning of global commerce, the freedom to practice technology and service standards, etc. The *quid pro quo* criterion must necessarily be a statistical criterion according to which some benefit, some do not, some are agnostic, and some are antagonistic, and it must be a dynamic criterion that is periodically reviewed in the "best interest" of humankind.
3. *Unenforceability* criterion. If a claimed element of a patented invention, in principle, can be performed wholly in the mind by any person, and this fact is known prior to patent grant, then the claim should not be allowed. If post-grant, it becomes known at any time that a claimed element of a patented invention can be performed wholly in the mind by a person, then the claim cannot thenceforth be infringed by any act that would otherwise have been considered infringing. For example, diagnostic tests may be patentable but not the mental diagnosis by a physician or anyone else or a similar diagnosis made by any other means. Since it is not possible to control peoples' thoughts, infringement of such claims cannot be controlled without violating a person's natural right to free thought. In this respect the patent is unenforceable.
4. *Scientific discovery and algorithm* criterion. Laws of nature are universally pervasive. These are Nature's prohibitory laws in that they will not allow any activity or creation that violate the laws. These laws primarily deal with matter; energy; space; time; motion; forces; transformations of matter and energy; and detection of natural phenomena by observers, sensors and detectors. They govern all human activity, including creating patentable inventions. Discovery of natural laws requires human ingenuity. However, humans can only know these laws not in their exact form but as guesses by making conjectures and refining them through diligent refutation of what has been discovered earlier. For this reason, discovery of a conjectured law of nature is patent eligible subject matter. Since laws of nature, as far as physicists can determine, are expressible in mathematical form (Tegmark), mathematical algorithms, whether known or newly discovered, when specifically interpreted (*i.e.*, given a specific meaning) by relating them to elements that constitute the universe or processes believed to be permitted to occur in the universe, are also patent eligible subject matter. A mathematical algorithm without an accompanying specific interpretation that connects it to the real world is not patentable subject matter for lack of utility. Interpretations not categorically claimed in a patent cannot be claimed under the doctrine of equivalents in patent enforcement.

5.5.1 Patent Validation Board

The legal validity of a granted patent's claims, if challenged, should not be decided by the courts but by an independent statutory body, which we here call the *Patent Validation Board*. The Board's decision shall not be contested in a court of law unless there is clear evidence of corrupt practices indulged by the Board that could have impacted the decision. On such evidence, the court shall have the patent re-examined by a new Board. The Board may ask the USPTO to re-examine and reissue an amended version of the patent, if feasible. A reissued patent shall be treated as a new patent for validation purposes. A patent needs to be validated only once by the Board; it can be done at any time during the patent's life and the validation may be requested by anyone. It would be in the interest of the patentee to have his patent validated before engaging in any licensing or other commercial activity or litigation.

The Board should comprise experts in patent examination, STEM experts, experts in patent law and members from the National Academies, all suitably chosen keeping the patented invention in mind. The Board should be supported by an expert prior art search team. The Board may also crowd-source to find prior art. The Board shall de novo determine the relevant PHOSITA for the patent. The first question it should settle before anything else is the *quid pro quo* aspect of patent grant: "Would society have benefited more if the patent had not been granted?" If the answer is yes, the patent should be revoked. The Board must decide keeping in mind the words of Thomas Jefferson:

> Considering the exclusive right to invention as given not of natural right, but for the benefit of society, I know well the difficulty of drawing a line between the things which are worth to the public the embarrassment of an exclusive patent, and those which are not.[74]

In litigation, the validity of the patent-in-suit must first be established by a Patent Validation Board, if not already done. For a valid patent, issues related to expansion of claim scope under the doctrine of equivalents or the applicability of the reverse doctrine of equivalents should be referred to the Patent Validation Board by the courts. The Board's decision in this respect shall be final.

5.6 Conclusions

In the confines of these pages arguments have been presented that compel us to believe that it is not just the patent system but the fate of the entire human race that is at stake. We the *Homo sapiens* may no longer exist a few centuries hence. The patent system encouraged and made it profitable for humans to use their ingenuity to innovate, share their innovations in a certain equitable way, and collectively improve the living conditions of fellow humans. Like all human constructed systems, it was never perfect, but it did wonders to give shape to our collective dreams and fantasies,

---

[74] Jefferson (1813).

e.g., visiting the planets, exploring space, global communication, flying in the air, and so on. Indeed, it has gone beyond our wildest dreams. It has now led to artificial intelligence, synthetic biology, and quantum computing. Collectively the trio have the unimaginable power to wipe us out of existence. The best the patent system can do is to adapt and reconfigure itself so that it can gracefully wind itself down before the *Homo sapiens* become extinct. There are many secrets about biological life that we do not know, e.g., there exist on Earth creatures with incredible superpowers of survival even after being frozen and suffocated, resist aging, regrow organs, defy cancer, etc. Any new understanding of these could revolutionize medicine, space travel and even war.[75] It is likely that our successor species may find and build their own world with this kind of new knowledge.

AI inventions by their very nature cannot be granted patent rights or such rights protected. AI inventions at their core belong to abstract mathematics and their most complex applications are essentially controlled by algorithms that are mechanizable computations. In fact, we already know some of the tricks that would allow AI machines to develop and discover new algorithms in a mechanized way. Human ingenuity may be rare, AI ingenuity will be pervasive.

References

Ainsworth C (2019) Want to regrow organs and defy cancer? Just copy these awesome animals. New Scientist

Ashton TS (1955) An economic history of England: the 18th century. Taylor and Francis, 1955, p. 107.

Bera RK (2009) The story of the cohen-boyer patents. Curr Sci 96(6):760–763. http://www.currentscience.ac.in/Downloads/article_id_096_06_0760_0763_0.pdf

Bera RK (2012) Biotechnology patents: safeguarding human health, in innovations in biotechnology, in Agbo EC (ed). InTech, ISBN 978–953–51–0096–6, Chapter 15, pp 349–376. http://www.intechopen.com/source/pdfs/28719/InTech-Biotechnology_patents_safeguarding_human_health.pdf. (Book is available at http://www.intechweb.org/books/show/title/innovations-in-biotechnology and can be downloaded for free)

Bera RK (2015) How valid are judicial exceptions in subject matter eligibility in U.S. patent law?. Available at SSRN: https://ssrn.com/abstract=2604737 or https://doi.org/10.2139/ssrn.2604737

Bera RK (2015a) Intellectual property rights: the new wealth of nations. Available at SSRN: http://ssrn.com/abstract=2572850 or https://doi.org/10.2139/ssrn.2572850

Bera RK (2015b) Synthetic biology and intellectual property rights. In Biotechnology, in: Ekinci D (ed). InTech, ISBN 978–953–51–2040–7, Chapter 9, pp 195–232. http://www.intechopen.com/download/pdf/48297. (Book is available at http://www.intechopen.com/books/biotechnology)

Bera RK (2015c) Rethinking patentable subject matter and related issues. Available at SSRN: https://ssrn.com/abstract=2699219 or https://doi.org/10.2139/ssrn.2699219

Bera RK (2016) Patent subject matter eligibility. Available at SSRN: https://ssrn.com/abstract=2883838 or https://doi.org/10.2139/ssrn.2883838

Bera RK (2018) AI powered society. Available at SSRN: https://ssrn.com/abstract=3256873. An updated version is available at Bera RK (2019) AI powered society. Adv Compu Commun, pp 59–85. https://acc.digital/aipowered-society/

---

[75] Ainsworth (2019).

Bera RK (2019) Synthetic biology, artificial intelligence, and quantum computing. book chapter in genetic engineering technology and synthetic biology, in: Nagpal ML (ed) IntechOpen, London, (to appear in 2019). Prepublication access to the chapter has been provided by InTech at https://www.intechopen.com/online-first/synthetic-biology-artificial-intelligence-and-quantum-computing; PDF at https://www.intechopen.com/chapter/pdf-download/65149

Bera RK, Menon V (2009) A new interpretation of superposition, entanglement, and measurement in quantum mechanics, arXiv:0908.0957v1[quant-ph]. http://arxiv.org/abs/0908.0957

Bernstein WJ (2004) The birth of plenty: how the modern world of prosperity was launched. McGraw-Hill

Brookes JC (2017) Quantum effects in biology: golden rule in enzymes, olfaction, photosynthesis and magnetodetection. Proc Royal Soc A Math Phys Eng Sci 473:20160822. https://doi.org/10.1098/rspa.2016.0822

Brooks M (2015) Is quantum physics behind your brain's ability to think? New Scientist. Available from: https://www.newscientist.com/article/mg22830500-300-is-quantum-physics-behind-your-brains-ability-to-think/ Accessed 02 Dec 2018

CAFC (2002) Johnson and Johnston Associates, Inc. v. RE Service Co., 285 F.3d 1046 (Fed. Cir. Mar. 28, 2002) (en banc, per curiam), http://digital-law-online.info/cases/62PQ2D1225.htm

Carnot S (1824) Reflections on the motive power of fire and on machines fitted to develop that power. Bachelier, Paris. http://www.thermohistory.org/carnot.pdf

Clausius R (1850) Über die bewegende Kraft der Wärme, Part I, Part II, Annalen der Physik, 79:368–397, 500–524. See English Translation: On the moving force of heat, and the law regarding the nature of heat itself which are deducible therefrom, Phil. Mag. (1851) 2:1–21, 102–119. https://archive.org/details/londonedinburghd02lond

Cohen SN, Chang ACY, Boyer HW, Helling RB (1973) Construction of biologically functional bacterial plasmids in vitro. Proc Natl Acad Sci USA 70(11):3240–3244. http://www.pnas.org/content/70/11/3240.full.pdf

Cong L et al (2013) Multiplex genome engineering using CRISPR/Cas systems. Science 339(6121):819–823. http://www.ncbi.nlm.nih.gov/pmc/articles/PMC3795411/

Cooke NJ, Hilton ML (eds) (2015) Enhancing the effectiveness of team science. NAP. http://download.nap.edu/cart/download.cgi?&record_id=19007

Darwin C (1859) On the origin of species by means of natural selection, 1st edn. John Murray, London. http://darwin-online.org.uk/content/frameset?itemID=F373&viewtype=text&pageseq=1

Descartes R (1637) Discourse on the method of reasoning well and seeking truth in the sciences. http://www.earlymoderntexts.com/pdfs/descartes1637.pdf

Deutsch D (1998) The Fabric of Reality. Penguin. Chapter 1, p. 2.

Erdős P, Rényi A (1960) On the evolution of random graphs. Publ Hungarian Acad Sci 5(1):17–61. http://snap.stanford.edu/class/cs224w-readings/erdos60random.pdf

Feynman R (1965) The character of physical law. Modern Library Edition, 1994, Originally published by BBC in 1965, and in paperback by MIT Press, 1967

Fisher MPA (2015) Quantum cognition: the possibility of processing with nuclear spins in the brain. Ann Phys 362:593–602. https://doi.org/10.1016/j.aop.2015.08.020

Galileo Galilei (1623) Il Saggiatore (in Italian) (Rome, 1623); The Assayer, English translation: Stillman Drake and C. D. O'Malley, in The Controversy on the Comets of 1618. University of Pennsylvania Press. http://www.princeton.edu/~hos/h291/assayer.htm

Gibson D et al (2010) Creation of a bacterial cell controlled by a chemically synthesized genome. Science 329(5987):52–56. http://adsabs.harvard.edu/abs/2010Sci...329...52G

Harding TS (1941) Exploitation of the creators. Philos Sci 8(3):385–390, at 386–387

Heathcotte B, Robert JS (2006) The strange case of the humanzee patent quest. Natl Catholic Bioethics Quarterly 6(1), Spring. https://www.academia.edu/12118574/The_Strange_Case_of_the_Humanzee_Patent_Quest

Hofstadter D (1979) Gödel, Escher, Bach: an eternal golden braid, Basic Books

Inkster I (2006) Potentially global: a story of useful and reliable knowledge and material progress in Europe, circa 1474–1914. Int History Rev XXVIII(2):237–286. http://www.jstor.org/discover/. https://doi.org/10.2307/40109747. An earlier version is available at http://grammatikhilfe.com/economicHistory/Research/GEHN/GEHNPDF/conf4_IInkster.pdf

Jefferson T (1813) Letter to Isaac McPherson. Monticello. In: The Letters of Thomas Jefferson. From Revolution to Reconstruction and Beyond. American History. http://www.let.rug.nl/usa/presidents/thomas-jefferson/letters-of-thomas-jefferson/jefl220.php

Landauer R (1961) Irreversibility and heat generation in the computing process. IBM J Research Dev, 5(3):183–191, Reprinted in IBM J Research Dev, 44(1–2):261–269, January/March 2000. http://www.pitt.edu/~jdnorton/lectures/Rotman_Summer_School_2013/thermo_computing_docs/Landauer_1961.pdf

Landauer R (1991) Information is physical. Phys Today, 23–29

Malyshev DA et al (2014) A semi-synthetic organism with an expanded genetic alphabet. Nature 509(7500):385–388. http://www.ncbi.nlm.nih.gov/pubmed/24805238

Maxwell JC (1865) A dynamical theory of the electromagnetic field. Philos Trans Royal Soc London 155:459–512. http://rstl.royalsocietypublishing.org/content/155/459. (This article accompanied a December 8, 1864 presentation by Maxwell to the Royal Society)

May RM (1976) Simple mathematical models with very complicated dynamics. Nature 261:459–467. http://abel.harvard.edu/archive/118r_spring_05/docs/may.pdf

Nelson DL, Cox MM (2006) Lehninger principles of biochemistry. 4th edn. https://www.academia.edu/4622632/Lehninger_Principles_of_Biochemistry_Fourth_Edition_-_David_L._Nelson_Michael_M._Cox

Newton I (1687) Philosophiæ naturalis principia mathematica. https://www.gutenberg.org/files/28233/28233-pdf.pdf

Nielsen MA, Chuang IL (2000) Quantum computation and quantum information. Cambridge University Press. [Errata at http://www.squint.org/qci/]

NRC (2012) Committee on the Mathematical Sciences in 2025; Board on mathematical sciences and their applications; Division on Engineering and physical sciences; National research council. Fueling innovation and discovery: The Mathematical Sciences in the 21st Century. NAP. https://www.nap.edu/download/13373

Olson S (ed) (2015) The past half century of engineering—and a look forward. NAP, 2015. https://doi.org/10.17226/21702.

Palmisano SJ (2003) The globally integrated enterprise. Foreign Affairs 85(3):127–136. http://www.oecd.org/sti/ieconomy/37998124.pdf

Popper K (1934) The Logic of scientific discovery. Routledge, 1959. (Originally published in German, Logik der Forschung, Mohr Siebeck, 1934)

Popper K (1963) Conjectures and refutations: the growth of scientific knowledge (Reprinted Ed., 2004). (First published, 1963), Routledge

Prather KLJ (2010) Applications of synthetic biology. Presidential Commission for the Study of Bioethical Issues. http://bioethics.gov/sites/default/files/Applications-of-Synthetic-Biology.ppt

SCOTUS (1950) Graver Tank and Mfg. Co. v. Linde Air Products, 339 U.S. 605. http://supreme.justia.com/us/339/605/case.html

SCOTUS (1961) Aro Mfg. v. Convertible Top Replacement Co., 365 U.S. 336, 339 (1961), http://supreme.justia.com/us/365/336/case.html

SCOTUS (1996) Markman v. Westview Instruments, Inc., 517 U.S. 370:371–73. http://straylight.law.cornell.edu/supct/html/95-26.ZO.html

SCOTUS (2002) Festo Corp. v. Shoketsu Kinzoku Kogyokabushiki Co., 535 U.S. 722 (2002), http://www.law.cornell.edu/supct/pdf/00-1543P.ZO (vacated and remanded)

SCOTUS (2013) Ass'n for Molecular Pathology v. Myriad Genetics, Inc., 133 S. Ct. 2107:2111. http://www.supremecourt.gov/opinions/12pdf/12-398_1b7d.pdf

SCOTUS (2014) Nautilus, Inc. v. BioSig Instruments, Inc., 572 U.S. https://supreme.justia.com/cases/federal/us/572/13-369/

Shannon CE (1948) A mathematical theory of communication. Bell System Technical Journal, 27, pp. 379–423 & 623–656, July & October, 1948, http://cm.bell-labs.com/cm/ms/what/shannonday/shannon1948.pdf

Shapiro Z (2015) Patent law, expertise, and the Court of Appeals for the Federal Circuit. Harvard Law. http://blogs.law.harvard.edu/billofhealth/2015/07/14/patent-law-expertise-and-the-court-of-appeals-for-the-federal-circuit/

Sharlach M (2014) CRISPR pioneers honored. The Scientist. http://www.the-scientist.com/?articles.view/articleNo/41455/title/CRISPR-Pioneers-Honored/

Stone P et al (2016) Artificial intelligence and life in 2030. One Hundred Year Study on Artificial Intelligence: Report of the 2015–2016 Study Panel (2016). https://ai100.stanford.edu/sites/default/files/ai_100_report_0831fnl.pdf

Tegmark M (2014) Our mathematical universe: my quest for the ultimate nature of reality. Knopf

Thomson W (1854) (Lord Kelvin). On the dynamical theory of heat. Part V. Thermo-electric currents. Trans Royal Soc Edinburgh 21 (part I):123. http://www.biodiversitylibrary.org/item/126375#page/147/mode/1up

Turing AM (1936) On computable numbers, with an application to the Entscheidungsproblem. Proc London Math Soc S2–42(1):230–265, 1936–1937. https://www.cs.virginia.edu/~robins/Turing_Paper_1936.pdf Correction at: Turing AM (1938) On computable numbers, with an application to the Entscheidungsproblem. A Correction S2–43(1):544–546. http://www.turingarchive.org/viewer/?id=466&title=02

von Neumann J (1954) In: Taub AH (ed) The role of mathematics in the sciences and in society (1954) an address to Princeton alumni, published in John von Neumann, Collected Works

Watson JD, Crick FHC (1953a) Molecular structure of nucleic acids. Nature 171(4356):737–738. http://www.nature.com/scitable/content/molecular-structure-of-nucleic-acids-a-structure-13997975

Watson JD, Crick FHC (1953b) Genetical implications of the structure of deoxyribonucleic acid. Nature, 171(4361):pp. 964–967. http://www.nature.com/nature/dna50/watsoncrick2.pdf

Weiß J, Schwietring T (2018) The power of language: a philosophical-sociological reflection. Goethe-Institut. http://www.goethe.de/lhr/prj/mac/msp/en1253450.htm

Wigner EP (1960) The unreasonable effectiveness of mathematics in the natural sciences. Richard Courant lecture in mathematical sciences delivered at New York University, May 11, 1959; published Communications on Pure and Applied Mathematics 13(1):1–14. http://www.maths.ed.ac.uk/~aar/papers/wigner.pdf

Yin H et al (2014) Genome editing with Cas9 in adult mice corrects a disease mutation and phenotype. Nat Biotechnol. Published online 30 March 2014 (Corrected 31 March 2014). http://www.nature.com/nbt/journal/vaop/ncurrent/full/nbt.2884.html

Zinkernagel H (2016) Niels Bohr on the wave function and the classical/quantum divide. Stud History Philos Modern Phys 53:9–19. https://arxiv.org/abs/1603.00353

Zinkernagel, Zinkernagel H (2015) Are we living in a quantum world? Bohr and quantum fundamentalism. In: Aaserud F, Kragh H, editors. One hundred years of the Bohr atom: Proceedings from a conference. Scientia Danica. Series M: Mathematica et physica, vol. 1. Copenhagen: Royal Danish Academy of Sciences and Letters; pp. 419–434. http://philsci-archive.pitt.edu/11785/1/BohrQuantumWorld.pdf

GPSR Compliance

The European Union's (EU) General Product Safety Regulation (GPSR) is a set of rules that requires consumer products to be safe and our obligations to ensure this.

If you have any concerns about our products, you can contact us on

ProductSafety@springernature.com

In case Publisher is established outside the EU, the EU authorized representative is:

Springer Nature Customer Service Center GmbH
Europaplatz 3
69115 Heidelberg, Germany